"Brilliant, provocative, opinionated, poetic, and inspiring."
—Rupert Sheldrake, author of *The Rebirth of Nature*

"Remarkable . . . There is a logic to McKenna's two principal arguments that is difficult to dismiss."
—*Los Angeles Times Book Review*

"McKenna's daring leaps of thought and radical vision force us to recognize our limited understanding of the legacy of drug use and its impact on our world. . . . [His] startling conclusions . . . will delight . . . and outrage."
—*Booklist*

"Terence McKenna is a wild man, probing the depths of the great mysteries of the mind. He asks big questions, he rides the Edge. He's an explorer, a circumnavigator of human consciousness. Terence is on to something."
—Mickey Hart, drummer for the Grateful Dead and author of *Drumming at the Edge of Magic*

"Terence McKenna, true bard of our psychedelic birthright, must have kissed the Blarney Stone. . . . I have gained more insight into the true roots of religious experience from his inspired rants than in all my years of book learning."
—Howard Rheingold, editor of *Whole Earth Review* and author of *Virtual Reality*

ALSO BY TERENCE McKENNA

THE INVISIBLE LANDSCAPE
with Dennis McKenna

PSILOCYBIN: THE MAGIC MUSHROOM GROWER'S GUIDE
with Dennis McKenna

THE ARCHAIC REVIVAL

TRUE HALLUCINATIONS

FOOD OF THE GODS

THE SEARCH FOR THE ORIGINAL TREE OF KNOWLEDGE

A RADICAL HISTORY OF PLANTS, DRUGS, AND HUMAN EVOLUTION

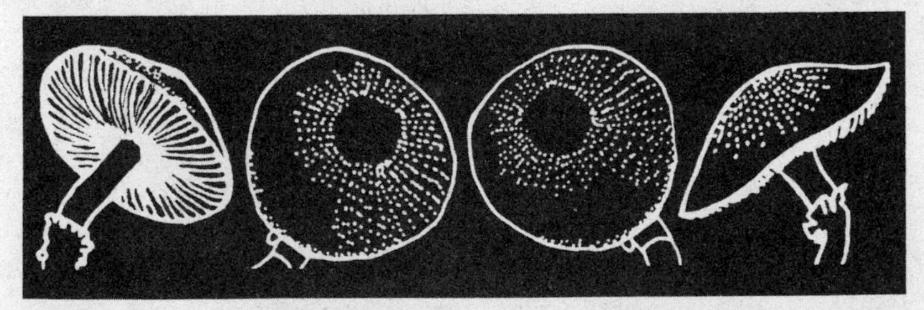

TERENCE McKENNA

BANTAM BOOKS
NEW YORK • TORONTO • LONDON • SYDNEY • AUCKLAND

FOOD OF THE GODS
A Bantam Book

PUBLISHING HISTORY
Bantam hardcover edition published March 1992
Bantam trade paperback edition/February 1993

See page 312 for acknowledgments.

Library of Congress Catalog Card Number 91-30534.

ISBN 978-0-553-37130-7

Published simultaneously
in the United States and Canada

Bantam Books are published by Bantam Books, a division of Bantam Doubleday Dell Publishing Group, Inc. Its trademark, consisting of the words "Bantam Books" and the portrayal of a rooster, is Registered in U.S. Patent and Trademark Office and in other countries. Marca Registrada. Bantam Books, New York, New York.

PRINTED IN THE UNITED STATES OF AMERICA

BVG 36 37 38 39

For

Kat

Finn and Klea

ACKNOWLEDGMENTS

I wish to thank my friends and colleagues for their patience and encouragement in the writing of this book, especially Ralph Abraham, Rupert Sheldrake, Ralph Metzner, Dennis McKenna, Chris Harrison, Neil Hassall, Dan Levy, Ernest Waugh, Richard Bird, Roy and Diane Tuckman, Faustin Bray and Brian Wallace, and Marion and Allan Hunt-Badiner. Thanks also to correspondents Dr. Elizabeth Judd and Marc Lamoreaux who passed along useful information. Each made their own unique contribution to my thinking, though my conclusions are mine to defend.

Archivist and friend Michael Horowitz made a deep contribution to this work. He read and criticized the manuscript carefully and made available the pictorial archives of the Fitz Hugh Ludlow Memorial Library, thus tremendously enriching the visual side of my argument. Thank you, Michael.

Special appreciation is offered to Michael and Dulce Murphy, Steve and Anita Donovan, Nancy Lunney, Paul Herbert, Kathleen O'Shaughnessy, and all of Esalen Institute for providing me with an opportunity to be the Esalen Scholar in Residence in June of 1989 and 1990. Parts of this book were written during those residencies. Thanks also to Lew and Jill Carlino and Robert Chartoff, patient friends who listened to parts of this book without, perhaps, realizing it.

My partner Kat, Kathleen Harrison McKenna, has long shared my passion for the psychedelic ocean and the ideas that swim there. In our voyages to the Amazon and elsewhere she has been the best possible companion, colleague, and muse.

Kat and my two children, Finn and Klea, supported me through writing this book, immune to my many moods and prolonged periods of writer's hibernation. To them I offer my deepest love and appreciation. Thanks for hanging in there, guys.

Very special thanks to Leslie Meredith, my editor at Bantam Books, and to her editorial assistant, Claudine Murphy. Their unflagging belief in the importance of these ideas was an inspiration to clarify and extend my thinking into new areas. Thanks also to my agent, John Brockman, who led me through the special initiation that only the Reality Club can give.

Lastly I want to acknowledge my deep debt to the psychedelic community, the hundreds of people that it has been my privilege to come into contact with during a lifetime spent in the pursuit of even a glimpse of the peacock angel. It is the shamans among us, both ancient and modern, those whose eyes have gazed on sights previously unseen by anyone, it is they who showed the way and who were my inspiration.

CONTENTS

INTRODUCTION: A MANIFESTO FOR NEW THOUGHT ABOUT DRUGS

A specter is haunting planetary culture—the specter of drugs. The definition of human dignity created by the Renaissance and elaborated into the democratic values of modern Western civilization seems on the point of dissolving. The major media inform us at high volume that the human capacity for obsessional behavior and addiction has made a satanic marriage with modern pharmacology, marketing, and high-speed transportation. Previously obscure forms of chemical use now freely compete in a largely unregulated global marketplace. Whole governments and nations in the Third World are held in thrall by legal and illegal commodities promoting obsessional behavior.

This situation is not new, but it is getting worse. Until quite recently international narcotics cartels were the obedient creations of governments and intelligence agencies that were searching for sources of "invisible" money with which to finance their own brand of institutionalized obsessional behavior.[1] Today, these drug cartels have evolved, through the unprecedented rise in the demand for cocaine, into rogue elephants before whose power even their creators have begun to grow uneasy.[2]

We are beset by the sad spectacle of "drug wars" waged by governmental institutions that usually are paralyzed by lethargy and inefficiency or are in transparent collusion with the international drug cartels they are publicly pledged to destroy.

No light can penetrate this situation of pandemic drug use and abuse unless we undertake a hard-eyed reappraisal of our present situation and an examination of some old, nearly forgotten, patterns of drug-related experience and behavior. The importance of this task cannot be overestimated. Clearly the self-administration of psychoactive substances, legal and illegal, will be increasingly a part of the future unfolding of global culture.

AN AGONIZING REAPPRAISAL

Any reappraisal of our use of substances must begin with the notion of habit, "a settled tendency or practice." Familiar, repetitious, and largely unexamined, habits are simply the things that we do. "People," says an old adage, "are creatures of habit." Culture is largely a matter of habit, learned from parents and those around us and then slowly modified by shifting conditions and inspired innovations.

Yet, however slow these cultural modifications may seem, when contrasted with the slower-than-glaciers modification of species and ecosystems, culture presents a spectacle of wild and continuous novelty. If nature represents a principle of economy, then culture surely must exemplify the principle of innovation through excess.

When habits consume us, when our devotion to them exceeds the culturally defined norms, we label them as obsessions. We feel, in such situations, as though the uniquely human dimension of free will has somehow been violated. We can become obsessed with almost anything: with a behavior pattern such as reading the morning paper or with material objects (the collector), land and property (the empire builder), or power over other people (the politician).

While many of us may be collectors, few of us have the opportunity to indulge our obsessions to the point of becoming empire builders or politicians. The obsessions of the ordinary person tend to focus on the here-and-now, on the realm of immediate gratification through sex, food, and drugs. An obsession with the chemical constituents of foods and drugs (also called metabolites) is labeled an addiction.

Addictions and obsessions are unique to human beings. Yes, ample anecdotal evidence supports the existence of a preference for intoxicated states among elephants, chimpanzees, and some butterflies.[3] But, as when we contrast the linguistic abilities of chimpanzees and dolphins with human speech, we see that these animal behaviors are enormously different from those of humans.

Habit. Obsession. Addiction. These words are signposts along a path of ever-decreasing free will. Denial of the power of free will is implicit in the notion of addiction, and in our culture, addictions are viewed seriously—especially exotic or unfamiliar addictions. In the nineteenth century the opium addict was the "opium fiend," a description that harkened back to the idea of a demonic possession by a controlling force from without. In the twentieth century, the addict as a person possessed has been replaced with the notion of addiction as disease. And, with the notion of addiction as disease, the role of free will is finally reduced to the vanishing point. After all, we are not responsible for the diseases that we may inherit or develop.

Today, however, human chemical dependence plays a more conscious role than ever before in the formation and maintenance of cultural values.

Since the middle of the nineteenth century and with ever-greater speed and efficiency, organic chemistry has placed into the hands of researchers, physicians, and ultimately everyone an endless cornucopia of synthetic drugs. These drugs are more powerful, more effective, of greater duration, and in some cases, many times more addictive than their natural relatives. (An exception is cocaine, which, although a natural product, when refined, concentrated, and injected is particularly destructive.)

The rise of a global information culture has led to the ubiquity of information on the recreational, aphrodisiacal, stimulating, sedative, and psychedelic plants that have been discovered by inquisitive human beings living in remote and previously unconnected parts of the planet. At the same time that this flood of botanical and ethnographic information arrived in Western society, grafting other cultures' habits onto our own and giving us greater choices than ever, great strides were being made in the synthesis of complex organic molecules and in the understanding of the molecular machinery of genes and heredity. These new insights and technologies are contributing to a very different culture of psychopharmacological engineering. Designer drugs such as MDMA, or Ecstasy, and anabolic steroids used by athletes and teenagers to stimulate muscle development are harbingers of an era of ever more frequent and effective pharmacological intervention in how we look, perform, and feel.

The notion of regulating, on a planetary scale, first hundreds and then thousands of easily produced, highly sought after, but illegal synthetic substances is appalling to anyone who hopes for a more open and less regimented future.

AN ARCHAIC REVIVAL

This book will explore the possibility of a revival of the Archaic—or preindustrial and preliterate—attitude toward community, substance use, and nature—an attitude that served our nomadic prehistoric ancestors long and well, before the rise of the current cultural style we call "Western." The Archaic refers to the Upper Paleolithic, a period seven to ten thousand years in the past, immediately preceding the invention and dissemination of agriculture. The Archaic was a time of nomadic pastoralism and partnership, a culture based on cattle-raising, shamanism, and Goddess worship.

I have organized the discussion in a roughly chronological order, with the last and most future-oriented sections taking up and recasting the Archaic themes of the early chapters. The argument proceeds along the lines of a pharmacological pilgrim's progress. Thus I have called the four sections of the book "Paradise," "Paradise Lost," "Hell," and, hopefully not too optimistically, "Paradise Regained?" A glossary of special terms appears at the end of the book.

Obviously, we cannot continue to think about drug use in the same old ways. As a global society, we must find a new guiding image for our culture, one that unifies the aspirations of humanity with the needs of the planet and the individual. Analysis of the existential incompleteness within us that drives us to form relationships of dependency and addiction with plants and drugs will show that at the dawn of history, we lost something precious, the absence of which has made us ill with narcissism. Only a recovery of the relationship that we evolved with nature through use of psychoactive plants before the fall into history can offer us hope of a humane and open-ended future.

Before we commit ourselves irrevocably to the chimera of a drug-free culture purchased at the price of a complete jettisoning of the ideals of a free and democratic planetary society, we must ask hard questions: Why, as a species, are we so fascinated by altered states of consciousness? What has been their impact on our esthetic and spiritual aspirations? What have we lost by denying the legitimacy of each individual's drive to use substances to experience personally the transcendental and the sacred? My hope is that answering these questions will force us to confront the consequences of denying nature's spiritual dimension, of seeing nature as nothing more than a "resource" to be fought over and plundered. Informed discussion of these issues will give no comfort to the control-obsessed, no comfort to know-nothing religious fundamentalism, no comfort to beige fascism of whatever form.

The question of how we, as a society and as individuals, relate to psychoactive plants in the late twentieth century, raises a larger question: how, over time, have we been shaped by the shifting alliances that we have formed and broken with various members of the vegetable world as we have made our way through the maze of history? This is a question that will occupy us in some detail in the chapters to come.

The Ur-myth of our culture opens in the Garden of Eden, with the eating of the fruit of the Tree of Knowledge. If we do not learn from our past, this story could end with a planet toxified, its forests a memory, its biological cohesion shattered, our birth legacy a weed-choked wasteland. If we have overlooked something in our previous attempts to understand our origins and place in nature, are we now in a position to look back and to understand, not only our past, but our future, in an entirely new way? If we can recover the lost sense of nature as a living mystery, we can be confident of new perspectives on the cultural adventure that surely must lie ahead. We have the opportunity to move away from the gloomy historical nihilism that characterizes the reign of our deeply patriarchal, dominator culture. We are in a position to regain the Archaic appreciation of our near-symbiotic relationship with psychoactive plants as a wellspring of insight and coordination flowing from the vegetable world to the human world.

The mystery of our own consciousness and powers of self-reflection is somehow linked to this channel of communication with the unseen mind that shamans insist is the spirit of the living world of nature. For shamans and shamanic cultures, exploration of this mystery has always been a credible alternative to living in a confining materialist culture. We of the industrial democracies can choose to explore these unfamiliar dimensions now or we can wait until the advancing destruction of the living planet makes all further exploration irrelevant.

A NEW MANIFESTO

The time has therefore come, in the great natural discourse that is the history of ideas, thoroughly to rethink our fascination with habitual use of psychoactive and physioactive plants. We have to learn from the excesses of the past, especially the 1960s, but we cannot simply advocate "Just say no" any more than we can advocate "Try it, you'll like it." Nor can we support a view that wishes to divide society into users and nonusers. We need a comprehensive approach to these questions that encompasses the deeper evolutionary and historical implications.

The mutation-inducing influence of diet on early humans and the effect of exotic metabolites on the evolution of their neurochemistry and culture is still unstudied territory. The early hominids' adoption of an omnivorous diet and their discovery of the power of certain plants were decisive factors in moving early humans out of the stream of animal evolution and into the fast-rising tide of language and culture. Our remote ancestors discovered that certain plants, when self-administered, suppress appetite, diminish pain, supply bursts of sudden energy, confer immunity against pathogens, and synergize cognitive activities. These discoveries set us on the long journey to self-reflection. Once we became tool-using omnivores, evolution itself changed from a process of slow modification of our physical form to a rapid definition of cultural forms by the elaboration of rituals, languages, writing, mnemonic skills, and technology.

These immense changes occurred largely as a result of the synergies between human beings and the various plants with which they interacted and coevolved. An honest appraisal of the impact of plants on the foundations of human institutions would find them to be absolutely primary. In the future, the application of botanically inspired steady-state solutions, such as zero population growth, hydrogen extraction from seawater, and massive recycling programs, may help reorganize our societies and planet along more holistic, environmentally aware, neo-Archaic lines.

The suppression of the natural human fascination with altered states of consciousness and the present perilous situation of all life on earth are intimately and causally connected. When we suppress access to shamanic ecstasy, we close off the refreshing waters of emotion that flow from having a deeply bonded, almost symbiotic relationship to the earth. As a consequence, the maladaptive social styles that encourage overpopulation, resource mismanagement, and environmental toxification develop and maintain themselves. No culture on earth is as heavily narcotized as the industrial West in terms of being inured to the consequences of maladaptive behavior. We pursue a business-as-usual attitude in a surreal atmosphere of mounting crises and irreconcilable contradictions.

As a species, we need to acknowledge the depth of our historical dilemma. We will continue to play with half a deck as long as we continue to tolerate cardinals of government and science who presume to dictate where human curiosity can legitimately focus its attention and where it cannot. Such restrictions on the human imagination are demeaning and preposterous. The government not only restricts research on psychedelics that could conceivably yield valuable psychological and medical insights, it presumes to prevent their religious and spiritual use, as well. Religious use of psychedelic plants is a civil rights issue; its restriction is the repression of a legitimate religious sensibility. In fact, it is not a religious sensibility that is being repressed, but *the* religious sensibility, an experience of *religio* based on the plant-human relationships that were in place long before the advent of history.

We can no longer postpone an honest reappraisal of the true costs and benefits of habitual use of plants and drugs versus the true costs and benefits of suppression of their use. Our global culture finds itself in danger of succumbing to an Orwellian effort to bludgeon the problem out of existence through military and police terrorism directed toward drug consumers in our own population and drug producers in the Third World. This repressive response is largely fueled by an unexamined fear that is the product of misinformation and historical ignorance.

Deep-seated cultural biases explain why the Western mind turns suddenly anxious and repressive on contemplating drugs. Substance-induced changes in consciousness dramatically reveal that our mental life has physical foundations. Psychoactive drugs thus challenge the Christian assumption of the inviolability and special ontological status of the soul. Similarly, they challenge the modern idea of the ego and its inviolability and control structures. In short, encounters with psychedelic plants throw into question the entire world view of the dominator culture.

We will come across this theme of the ego and the dominator culture often in this reexamination of history. In fact, the terror the ego feels in contemplating the dissolution of boundaries between self and world not only lies behind the suppression of altered states of consciousness but, more generally, explains the suppression of the feminine, the foreign and exotic, and transcendental experiences. In the prehistoric but post-Archaic times of about 5000 to 3000 B.C., suppression of partnership society by patriarchal invaders set the stage for suppression of the open-ended experimental investigation of nature carried on by shamans. In highly organized societies that Archaic tradition was replaced by one of dogma, priestcraft, patriarchy, warfare and, eventually, "rational and scientific" or dominator values.

To this point I have used the terms "partnership" and "dominator" styles of culture without explanation. I owe these useful terms to Riane Eisler and her important re-visioning of history, *The Chalice and the Blade.*[4] Eisler has advanced the notion that "partnership" models of society preceded and later competed with, and were oppressed by, "dominator" forms of social organization. Dominator cultures are hierarchical, paternalistic, materialistic, and male-dominated. Eisler believes that the tension between the partnership and dominator organizations and the overexpression of the dominator model are responsible for our alienation from nature, from ourselves, and from each other.

Eisler has written a brilliant synthesis of the emergence of human culture in the ancient Near East and the unfolding political debate concerning the feminizing of culture and the need to overcome patterns of male dominance in creating a viable future. Her analysis of gender politics raises the level of debate beyond those who have so shrilly hailed and decried this or that ancient "matriarchy" or "patriarchy." *The Chalice and the Blade* introduces the notion of "partnership societies" and "dominator societies" and uses the archaeological record to argue that over vast areas and for many centuries the partnership societies of the ancient Middle East were without warfare and upheaval. Warfare and patriarchy arrived with the appearance of dominator values.

THE DOMINATOR INHERITANCE

Our culture, self-toxified by the poisonous by-products of technology and egocentric ideology, is the unhappy inheritor of the dominator attitude that alteration of consciousness by the use of plants or substances is somehow wrong, onanistic, and perversely antisocial. I will argue that suppression of shamanic gnosis, with its reliance and insistence on ecstatic dissolution of the ego, has robbed us of life's meaning and made us enemies of the planet, of ourselves, and our grandchildren. We are killing the planet in order to keep intact the wrongheaded assumptions of the ego-dominator cultural style.

It is time for change.

I

PARADISE

1

SHAMANISM: SETTING THE STAGE

Raongi sat still in the fading light of the fire. He felt his body flex deep within in ways that reminded him of the gulping of an eel. As he formed this thought, an eel's head, oversized and bathed in electric blue, appeared obediently in the darkened space behind his eyelids.

"Mother spirit of the first waterfall . . ."

"Grandmother of the first rivers . . ."

"Show yourself, show yourself."

Responding to the voices, the darkened space behind the now slowly spinning eel apparition filled with sparks; waves of light leaped higher and higher, accompanied by a roar of increasing intensity.

"It is the first *maria*." The voice is that of Mangi, the elder shaman of the village of Jarocamena. "It is strong. So strong."

Mangi is silent as the visions close over them. They are on the brink of Ventúri, the real world, the blue zone. The sound of falling rain outside is unrecognizable. There is the shuffling of dry leaves mingled with the sound of distant bells. Their tingling seems more like light than sound.

. . .

Until relatively recently, the practices of Mangi and her remote Amazonian tribe were typical of religious practice everywhere. Only in the last several millennia have theology and ritual graduated to more elaborate—and not necessarily more serviceable—forms.

SHAMANISM AND ORDINARY RELIGION

When I arrived in the Upper Amazon in early 1970, I had just spent several years living in Asian societies. Asia is a place where the shattered shells of castoff religious ontologies litter the dusty landscape like the carapaces of sand-scoured scarabs. I had traveled India in search of the miraculous. I had visited its temples and ashrams, its jungles and mountain retreats. But Yoga, a lifetime calling, the obsession of a disciplined and ascetic few, was not sufficient to carry me to the inner landscapes that I sought.

I learned in India that religion, in all times and places where the luminous flame of the spirit has guttered low, is no more than a hustle. Religion in India stares from world-weary eyes familiar with four millennia of priestcraft. Modern Hindu India to me was both an antithesis and a fitting prelude to the nearly archaic shamanism that I found in the lower Rio Putumayo of Colombia when I arrived there to begin studying the shamanic use of hallucinogenic plants.

Shamanism is the practice of the Upper Paleolithic tradition of healing, divination, and theatrical performance based on natural magic developed ten to fifty thousand years ago. Mircea Eliade, author of *Shamanism: Archaic Techniques of Ecstasy* and the foremost authority on shamanism in the context of comparative religion, has shown that in all times and places shamanism maintains a surprising internal coherency of practice and belief. Whether the shaman is an Arctic-dwelling Inuit or a Witoto of the Upper Amazon, certain techniques and expectations remain the same. Most important of these invariants is ecstasy, a point my brother and I make in our book *The Invisible Landscape:*

> The ecstatic part of the shaman's initiation is harder to analyze, for it is dependent on a certain receptivity to states of trance and ecstasy on the part of the novice; he may be moody, somewhat frail and sickly, predisposed to solitude, and may perhaps have fits of epilepsy or catatonia, or some

> other psychological aberrance (though not always as some writers on the subject have asserted).[1] In any case, his psychological predisposition to ecstasy forms only the starting point for his initiation: the novice, after a history of psychosomatic illness or psychological aberration that may be more or less intense, will at last begin to undergo initiatory sickness and trances; he will lie as though dead or in deep trance for days on end. During this time, he is approached in dreams by his helping spirits, and may receive instruction from them. Invariably during this prolonged trance the novice will undergo an episode of mystical death and resurrection; he may see himself reduced to a skeleton and then clothed with new flesh; or he may see himself boiled in a caldron, devoured by the spirits, and then made whole again; or he may imagine himself being operated on by the spirits, his organs removed and replaced with "magical stones" and then sewn up again.

Eliade showed that, while the particular motifs may vary between cultures and even individuals, shamanism's general structure is clear: the neophyte shaman undergoes a symbolic death and resurrection, which is understood as a radical transformation into a superhuman condition. Henceforth, the shaman has access to the superhuman plane, is a master of ecstasy, can travel in the spirit realm at will, and most important, can cure and divine. As we noted in *The Invisible Landscape:*

> In short, the shaman is transformed from a profane into a sacred state of being. Not only has he effected his own cure through this mystical transmutation, he is now invested with the power of the sacred, and hence can cure others as well. It is of the first order of importance to remember this, that the shaman is more than merely a sick man, or a madman; he is a sick man who has healed himself, who is cured, and who must shamanize in order to remain cured.[2]

It should be noted that Eliade used the word "profane" deliberately with the intent of creating a clear split between the notion of the profane world of ordinary experience and the sacred world which is "Wholly Other."[3]

THE TECHNIQUES OF ECSTASY

Not all shamans use intoxication with plants to obtain ecstasy, but all shamanic practice aims to give rise to ecstasy. Drumming, manipulation of breath, ordeals, fasting, theatrical illusions, sexual abstinence—all are time-honored methods for entering into the trance necessary for shamanic work. Yet none of these methods is as effective, as ancient, and as overwhelming as the use of plants containing chemical compounds that produce visions.

This practice of using visionary plant intoxicants may seem alien or surprising to some Westerners. Our society regards psychoactive drugs as either frivolous or dangerous, at best reserved for the treatment of the seriously mentally ill where no other effective method is available. We retain the notion of the healer in the figure of the medical professional who, through the possession of special knowledge, can cure. But the specialized knowledge of the modern physician is clinical knowledge, removed from the unfolding drama of each unique and particular person.

Shamanism is different. Usually, if drugs are used, the shaman, not the patient, will take the drug. The motivation is also entirely different. The plants used by the shaman are not intended to stimulate the immune system or the body's other natural defenses against disease. Rather, the shamanic plants allow the healer to journey into an invisible realm in which the causality of the ordinary world is replaced with the rationale of natural magic. In this realm, language, ideas, and meaning have greater power than cause and effect. Sympathies, resonances, intentions, and personal will are linguistically magnified through poetic rhetoric. The imagination is invoked and sometimes its forms are beheld visibly. Within the magical mind-set of the shaman, the ordinary connections of the world and what we call natural laws are deemphasized or ignored.

A WORLD MADE OF LANGUAGE

The evidence gathered from millennia of shamanic experience argues that the world is actually made of language in some fashion. Although at odds with the expectations of modern science, this radical proposition is in agreement with much of current linguistic thinking.

"The twentieth-century linguistic revolution," says Boston University anthropologist Misia Landau, "is the recognition that language is not merely a device for communicating ideas about the world, but rather a tool for bringing the world into existence in the first place. Reality is not simply 'experienced' or 'reflected' in language, but instead is actually produced by language."[4]

From the point of view of the psychedelic shaman, the world appears to be more in the nature of an utterance or a tale than in any way related to the leptons and baryons or charge and spin that our high priests, the physicists, speak of. For the shaman, the cosmos is a tale that becomes true as it is told and as it tells itself. This perspective implies that human imagination can seize the tiller of being in the world. Freedom, personal responsibility, and a humbling awareness of the true size and intelligence of the world combine in this point of view to make it a fitting basis for living an authentic neo-Archaic life. A reverence for and an immersion in the powers of language and communication are the basis of the shamanic path.

This is why the shaman is the remote ancestor of the poet and artist. Our need to feel part of the world seems to demand that we express ourselves through creative activity. The ultimate wellsprings of this creativity are hidden in the mystery of language. Shamanic ecstasy is an act of surrender that authenticates both the individual self and that which is surrendered to, the mystery of being. Because our maps of reality are determined by our present circumstances, we tend to lose awareness of the larger patterns of time and space. Only by gaining access to the Transcendent Other can those patterns of time and space and our role in them be glimpsed. Shamanism strives for this higher point of view, which is achieved through a feat of linguistic prowess. A shaman is one who has attained a vision of the beginnings and the endings of all things *and who can communicate that vision*. To the rational thinker, this is inconceivable, yet the techniques of shamanism are directed toward this end and this is the source of their power. Preeminent among the shaman's techniques is the use of the plant hallucinogens, repositories of living vegetable gnosis that lie, now nearly forgotten, in our ancient past.

By entering the domain of plant intelligence, the shaman becomes, in a way, privileged to a higher dimensional perspective on experience. Common sense assumes that, though languages are always evolving, the raw stuff of what language expresses is relatively constant and common to all humans. Yet we also know that the Hopi language has no past or future tenses or concepts. How, then, can the Hopi world be like ours? And the Inuit have no first-person pronoun. How, then, can their world be like ours?

The grammars of languages—their internal rules—have been carefully studied. Yet too little attention has been devoted to examining how language creates and defines the limits of reality. Perhaps language is more properly understood when thought of as magic, for it is the implicit position of magic that the world is made of language.

If language is accepted as the primary datum of knowing, then we in the West have been sadly misled. Only shamanic approaches will be able to give us answers to the questions we find most interesting: who are we, where did we come from, and toward what fate do we move? These questions have never been more important than today, when evidence of the failure of science to nurture the soul of humanity is everywhere around us. Ours is not merely temporary ennui of the spirit; if we are not careful, ours is a terminal condition of the collective body *and* spirit.

The rational, mechanistic, antispiritual bias of our own culture has made it impossible for us to appreciate the mind-set of the shaman. We are culturally and linguistically blind to the world of forces and interconnections clearly visible to those who have retained the Archaic relationship to nature.

Of course, when I arrived in the Amazon twenty years ago, I knew nothing of the above. Like most Westerners, I believed that magic was a phenomenon of the naive and the primitive, that science could provide an explanation for the workings of the world. In that position of intellectual naïveté, I encountered psilocybin mushrooms for the first time, at San Augustine in the Alto Magdalena of southern Colombia. Later and not far away, in Florencia, I also encountered and used visionary brews made from *Banisteriopsis* vines, the yagé or *ayahuasca* of 1960s underground legend.[5]

The experiences that I had during those travels were personally transforming and, more important, they introduced me to a class of experiences that is vital to the restoration of balance in our social and environmental worlds.

I have shared the group mind that is generated in the vision sessions of the *ayahuasqueros*. I have seen the magical darts of red light that one shaman can send against another. But more revelatory than the paranormal feats of gifted magicians and spiritual healers were the inner riches that I discovered within my own mind at the apex of these experiences. I offer my account as a kind of witness, an Everyman; if these experiences happened to me, then they can be part of the general experience of men and women everywhere.

A SHAMANIC MEME

My shamanic education was not unique. Thousands of people have, by one means or another, come to the conclusion that psychedelic plants, and the shamanic institutions that their use implies, are profound tools for the exploration of the inner depths of the human psyche. Psychedelic shamans now constitute a worldwide and growing subculture of hyperdimensional explorers, many of whom are scientifically sophisticated. A landscape is coming into focus, a region still glimpsed only dimly, but emerging, claiming the attention of rational discourse—and possibly threatening to confound it. We may yet remember how to behave, how to take our correct place in the connecting pattern, the seamless web of all things.

An understanding of how to achieve this balance lingers on in the forgotten and trampled cultures of the rain forests and deserts of the Third World, and in the reserves and reservations into which dominator cultures force their aboriginal people. The shamanic gnosis is possibly dying; certainly it is changing. Yet the plant hallucinogens that are the source of this, the oldest of human religions, remain a clear running spring, as refreshing as they have always been. Shamanism is vital and real because of the individual encounter with the challenge and wonder, the ecstasy and exaltation induced by hallucinogenic plants.

My encounters with shamanism and hallucinogens in the Amazon convinced me of their salvific importance. Once convinced,

I was determined to filter out the various forms of linguistic, cultural, pharmacological, and personal noise that obscured the Mystery. I hoped to distill the essence of shamanism, to track the Epiphany to her lair. I wanted to see beyond the veils of her whirling dance. A cosmic peeping Tom, I dreamed of confronting naked beauty.

A cynic in the dominator style might be content to dismiss this as delusions of romantic youth. Ironically, I was at one time that cynic. I felt the folly of the quest. I knew the odds. "The Other? Naked Platonic beauty? You must be kidding!"

And it must be admitted that there were many wild misadventures along the way. "We must become God's fools," an enthusiastic Zen acquaintance once urged, by which he meant, "Hit the road." Seeking and finding had been a method that had worked for me in the past. I knew that shamanic practices based on the use of hallucinogenic plants still survived in the Amazon, and I was determined to confirm my intuition that a great secret lay undiscovered behind this fact.

• • •

Reality outran apprehension. The mottled face of the leprous old woman was made more startlingly hideous when the fire she tended suddenly flared as she added more wood. In the semidarkness behind her, I could see the guide who had brought me to this unnamed place on the Rio Cumala. Back in the river town bar this chance encounter with a boatman willing to take me to see the miracle-working *ayahuasca* witch of local legend had seemed like a great chance for a story. Now, after three days of river travel and a half-day struggling through trails so flooded with mud that they threatened with each step to suck your boots right off you, I was not so sure.

At this point, the original object of my quest—the authentic deep forest *ayahuasca*, reportedly so different from the swill of the charlatans of the marketplace—hardly held any interest for me.

"*Tomé, caballero!*" the old woman had cackled as she offered me a full cup of the black, slow-flowing liquid. Its surface had the sheen of motor oil.

She must have grown into this role, I thought as I drank. It was warm and salty, chalky and bittersweet. It tasted like the blood of some old, old thing. I tried not to think about how much at the mercy of these strange people I now was. But in fact my courage was failing. Both Doña Catalina and the guide's mocking eyes had slowly gone cold and mantislike. A wave of insect sound sweeping up the river seemed to splatter the darkness with shards of sharp-edged light. I felt my lips go numb.

Trying not to appear as loaded as I felt, I crossed to my hammock and lay back. Behind my closed eyelids there was a flowing river of magenta light. It occurred to me in a kind of dreamy mental pirouette that a helicopter must be landing on top of the hut, and this was the last impression I had.

When I regained consciousness I appeared to myself to be surfing on the inner curl of a wave of brightly lit transparent information several hundred feet high. Exhilaration gave way to terror as I realized that my wave was speeding toward a rocky coastline. Everything disappeared in the roaring chaos of informational wave meeting virtual land. More lost time and then an impression of being a shipwrecked sailor washed onto a tropical shore. I feel that I am pressing my face into the hot sand of a tropical beach. I feel lucky to be alive. I *am* lucky to be alive! Or is it that I am alive to be lucky? I break up laughing.

At this point the old woman begins to sing. Hers is no ordinary song, but an *icaro*, a magical curing song that in our intoxicated and ecstatic state seems more like a tropical reef fish or an animated silk scarf of many colors than a vocal performance. The song is a visible manifestation of power, enfolding us and making us secure.

SHAMANISM AND THE LOST ARCHAIC WORLD

Shamanism was beautifully defined by Mircea Eliade as "the archaic techniques of ecstasy." Eliade's use of the term "archaic" is important here because it alerts us to the role that shamanism must play in any authentic revival of vital Archaic forms of being, living, and understanding. The shaman gains entrance into a world that is hidden from those who dwell in ordinary reality. In this other dimension lurk powers both helpful and malevolent. Its rules are not the rules of our world; they are more like the rules that operate in myth and dream.

Shamanic healers insist on the existence of an intelligent Other somewhere in a dimension nearby. The existence of an ecology of souls or a disincarnate intelligence is not something that science can be expected to grapple with and emerge with its own premises intact. Particularly if this Other has long been a part of the terrestrial ecology, present but unseen, a global secret sharer.

The writings of Carlos Castaneda and his imitators have resulted in a fad of "shamanic awareness" which, although muddled, has turned the shaman from a peripheral figure in the literature of cultural anthropology into *the* media role model for full membership in neo-Archaic society. In spite of the grip that shamanism has on the popular imagination, the paranormal phenomena that it assumes to be actual and true have never been taken seriously by modern science, even if scientists, in a rare case of deference, have called on psychologists and anthropologists to analyze shamanism. This blindness to the paranormal world has created an intellectual blind spot in our normal world view. We are completely unaware of the magical world of the shaman. It is quite simply stranger than we *can* suppose.

Consider a shaman who uses plants to converse with an invisible world inhabited by nonhuman intelligences. This would seem to rate a headline in the supermarket tabloids. Yet anthropologists report such things all the time and no one raises an eyebrow. That is because we tend to assume that the shaman *interprets* his experience of intoxication as communication with spirits or ancestors. The implication is that you or I would interpret this same experience differently and that therefore it is no big deal that some poor, uneducated *campesino* thought he was talking with an angel.

Xenophobic as this attitude is, it suggests a good operating procedure since what it is saying is, "Show me the techniques of your ecstasy and I will judge their effectiveness for myself." I did this. This is my credential for the theories and opinions I hold. I was initially appalled at what I found: the world of shamanism, of allies, shape shifting, and magical attack are far more real than the constructs of science can ever be, because these spirit ancestors and their other world can be seen and felt, they can be known, in the nonordinary reality.

Something profound, unexpected, nearly unimaginable awaits us if we will turn our investigative attentions toward the phenomenon of shamanic plant hallucinogens. The people outside of Western history, those still in the dream time of preliteracy, have kept the flame of a tremendous mystery burning. It will be humbling to admit this and to learn from them, but that too is a part of the Archaic Revival.

This is not to imply that we must stand slack-jawed before the accomplishments of the "primitive" in yet another version of the Noble Savage Cha-Cha. Everyone who has worked in the field is aware of the frequent clash between our expectations of how "true rain forest people" should behave and the realities of tribal daily life. No one yet understands the mysterious intelligence within plants or the implications of the idea that nature communicates in a basic chemical language that is unconscious but profound. We do not yet understand how hallucinogens transform the message in the unconscious into revelations beheld by the conscious mind. As archaic people honed their intuitions and their senses by using whatever plants were at hand to increase their adaptive advantage, they had little time for philosophy. To this day the implications of the existence of this mind within nature discovered by shamanic peoples have yet to fully dawn.

Meanwhile, quietly and outside of history, shamanism has pursued its dialogue with an invisible world. Shamanism's legacy can act as a steadying force to redirect our awareness toward the collective fate of the biosphere. The shamanic faith is that humanity is not without allies. There are forces friendly to our struggle to birth ourselves as an intelligent species. But they are quiet and shy; they are to be sought, not in the arrival of alien star fleets in the skies of earth, but nearby, in wilderness solitude, in the ambience of waterfalls, and yes, in the grasslands and pastures now too rarely beneath our feet.

2

THE MAGIC IN FOOD

For days the Fox Clan had been gathering and storing unusually large amounts of food. Strips of gazelle meat had been smoked to a uniform darkness, while the clan children had gathered sweet grass corms and insect pupas. And the women had amassed eggs—the largest number ever. These eggs preoccupied Lami, who was careful to attend to the task before her. After all, was she not the daughter of the Mistress of All Birds? The eggs had to be carefully stacked in open wicker baskets and transported on the heads of some of the more responsible girls. The food trading ritual would take place when the people of the Fox Clan, Lami's people, met the Hawk people, the mysterious dwellers in the land of sandstone pinnacles. This very day they would join those others, as they had each year for time untellable, for the great festival dances and the food exchange. Lami recalled the last gathering of her sib, when Venda, the most-cycles shaman of the Fox People, had proclaimed the festival and its motivation.

"To share food is to be of one body. As the Hawk Clan eats of our food they become as we are. As we eat of their food we become them. Through eating the food of the others, we remain as one." With her shriveled breasts and bent back, Venda seemed ancient to Lami. Whatever her age, no one remembered more than she, and her word was rarely questioned within the group. Lami gently hoisted her burden for the trek. If the Hawk people wanted eggs, then eggs they would have.

. . .

The ways in which humans use plants, foods, and drugs cause the values of individuals and, ultimately, whole societies to shift. Eating some foods makes us happy, eating others sleepy, and still others alert. We are jovial, restless, aroused, or depressed depending on what we have eaten. Society tacitly encourages certain behaviors that correspond to internal feelings, thereby encouraging the use of substances that produce acceptable behaviors.

Suppression or expression of sexuality, fertility and sexual potency, degree of visual acuity, sensitivity to sound, speed of motor response, rate of maturation, and lifespan—these are only some of an animal's characteristics that can be influenced by food plants with exotic chemistries. Human symbol formation, linguistic facility, and sensitivity to community values may also shift under the influence of psychoactive and physiologically active metabolites. A night spent observing behavior in a singles bar should be fieldwork enough to confirm this observation. Indeed the mate-getting hustle has always placed a high premium on linguistic facility, as perennial attention to patter styles and opening lines attests.

When thinking about drugs, we tend to focus on episodes of intoxication, but many drugs are normally used in subthreshold or maintenance doses; coffee and tobacco are obvious examples in our culture. The result of this is a kind of "ambience of intoxication." Like fish in water, people in a culture swim in the virtually invisible medium of culturally sanctioned yet artificial states of mind.

Languages appear invisible to the people who speak them, yet they create the fabric of reality for their users. The problem of mistaking language for reality in the everyday world is only too well known. Plant use is an example of a complex language of chemical and social interactions. Yet most of us are unaware of the effects of plants on ourselves and our reality, partly because we have forgotten that plants have always mediated the human cultural relationship to the world at large.

A SHAGGY PRIMATE STORY

At Gombe Stream National Park in Tanzania, primatologists found that one particular species of leaf kept appearing undigested in chimpanzee dung. They found that every few days the chimps, instead of eating wild fruit as usual, would walk for twenty minutes or more to a site where a species of *Aspilia* grew. The chimps would repeatedly place their lips over an *Aspilia* leaf and hold it in their mouths. They would pluck a leaf, place it in their mouths, roll it around for a few moments, then swallow it whole. In this way as many as thirty small leaves might be eaten.

Biochemist Eloy Rodriguez of the University of California at Irvine isolated the active principle from the *Aspilia*—a reddish oil now named thiarubrine-A. Neil Towers of the University of British Columbia found that this compound can kill common bacteria in concentrations of less than one part per million. Herbarium records studied by Rodriguez and Towers showed that African peoples used *Aspilia* leaves to treat wounds and stomachaches. Of the four species native to Africa, the indigenous peoples used only three, the same three species used by the chimpanzees.[1]

Rodriguez and Towers have continued their observations of chimp and plant interactions and can now identify nearly a dozen plants, a veritable materia medica, in use among chimpanzee populations.

YOU ARE WHAT YOU EAT

Our proposed story of human emergence into the light of self-reflection is a you-are-what-you-eat story. Major climatic change and a newly broadened and hence mutagenic diet provided many opportunities for natural selection to affect the evolution of major human traits. Each encounter with a new food, drug, or flavoring was fraught with risk and unpredictable consequences. And this is even more true today, when our food contains hundreds of poorly studied preservatives and additives.

As an example of plants with a potential impact on a human population, consider sweet potatoes of the genus *Dioscorea*. In much of the tropical world, sweet potatoes provide a reliable and nutritious source of food. Nevertheless, several closely related species contain compounds that can interfere with ovulation. (These have become the source of raw materials for modern birth control pills.) Something close to genetic chaos would descend on a population of primates that settled into feeding upon these species of *Dioscorea*. Many such scenarios, though of a less spectacular magnitude, must have occurred as early hominids experimented with new foods while expanding their omnivorous dietary habits.

Eating a plant or an animal is a way of claiming its power, a way of assimilating its magic to one's self. In the minds of preliterate people, the lines between drugs, foods, and spices are rarely clearly drawn. The shaman who gorges himself on chili peppers to raise inner heat is hardly in a less altered state than the nitrous oxide enthusiast after a long inhalation. In our perception of flavor and our pursuit of variety in the sensation of eating, we are markedly different from even our primate cousins. Somewhere along the line, our new omnivorous eating habits and our evolving brain with its capacity to process sensory data were united in the happy notion that food can be experience. Gastronomy was born—born to join pharmacology, which must surely have preceded it, since maintenance of health through regulation of diet is seen among many animals.

The strategy of the early hominid omnivores was to eat everything that seemed foodlike and to vomit whatever was unpalatable. Plants, insects, and small animals found edible by this method were then inculcated into their diet. A changing diet or an omnivorous diet means exposure to an ever-shifting chemical equilibrium. An organism may regulate this chemical input through internal processes but, ultimately, mutagenic influences will increase and a greater than usual number of genetically variant individuals will be offered up to the process of natural selection. The results of this natural selection are accelerated changes in neural organization, states of consciousness, and behavior. No change is permanent, each gives way to yet another. All flows.

SYMBIOSIS

As plants influenced the development of humans and other animals, so were the plants themselves affected in turn. This coevolution invokes the idea of symbiosis. "Symbiosis" has several meanings; I use it to mean a relationship between two species that confers mutual benefit upon their members. The biological and evolutionary success of each species is linked to and enhanced by that of the other. This situation is the opposite of parasitism, though happy is the parasite that can evolve into a symbiont. Symbiotic relationships, in which each member requires the other, can be very tightly bound genetically or the linkage can be somewhat more open. While human-plant interactions were symbiotic in their pattern of mutual gain and advantage, these relationships were not genetically programmed. They are clearly seen instead as deep habits when contrasted with examples of true symbiosis from the world of nature.

One example of a genetically bound and hence truly symbiotic relationship involves the small clown anemone fish, *Amphiprion ocellaris*, which spends its life in the proximity of certain species of sea anemones. These clown fish are protected from larger predators by the anemones, and the anemones' food supply is expanded by the clown fish, which attract larger fish into the area where the anemone is feeding. When such a mutually agreeable arrangement is in place for a long time, it will eventually "institutionalize" itself by progressively blurring the clear genetic distinction between the symbionts. Ultimately one organism may actually become a part of the other, much in the way that mitochondria, the powerhouses of the animal cell, joined with other structures to form the cell. Mitochondria have a separate genetic component, whose ancestry can be traced to free-swimming eukaryotic bacteria, which once, hundreds of millions of years ago, were independent organisms.

Another instance of symbiosis that is instructive and that may have deep implications for our own situation is the relationship that has evolved between leafcutter ants and a species of *basidiomycete*, a mushroom. E. O. Wilson discusses this relationship:

> At the end of the trail the burdened foragers rush down the nest hole, into throngs of nestmates and along tortuous channels that end near the water table fifteen feet or more below. The ants drop the leaf sections onto the floor of a chamber, to be picked up by workers of a slightly smaller size who clip them into fragments about a millimeter across. Within minutes still smaller ants take over, crush and mold

the fragments into moist pellets, and carefully insert them into a mass of similar material. This mass ranges in size between a clenched fist and a human head, is riddled with channels, and resembles a gray cleaning sponge. It is the garden of the ants: on its surface a symbiotic fungus grows which, along with the leaf sap, forms the ants' sole nourishment. The fungus spreads like a white frost, sinking its *hyphae* into the leaf paste to digest the abundant cellulose and proteins held there in partial solution.

The gardening cycle proceeds. Worker ants even smaller than those just described pluck loose those strands of the fungus from places of dense growth and plant them onto the newly constructed surfaces. Finally, the very smallest—and most abundant—workers patrol the beds of fungal strands, delicately probing them with their antennae, licking their surfaces clean, and plucking out the spores and *hyphae* of alien species of mold. These colony dwarfs are able to travel through the narrowest channels deep within the garden masses. From time to time they pull tufts of fungus loose and carry them out to their larger nestmates.

No other animals have evolved the ability to turn fresh vegetation into mushrooms. The evolutionary event occurred only once, millions of years ago, somewhere in South America. It gave the ants an enormous advantage: They could now send out specialized workers to collect the vegetation while keeping the bulk of their populations safe in subterranean retreats. As a result, all of the different kinds of leafcutters together, comprising fourteen species in the genus *Atta* and twenty-three in *Acromyrmex*, dominate a large part of the American tropics. They consume more vegetation than any other group of animals, including the more abundant forms of caterpillars, grasshoppers, birds, and mammals.[2]

We can forgive E. O. Wilson, the foremost exponent of sociobiology, for thinking that an animal and a mushroom formed a mutually beneficial relationship only once in the history of the earth. His description of leafcutter ant society and its relationship to fungal agriculture anticipates and introduces central considerations in my effort to re-vision our own complex relationship to plants. For as we shall see, a by-product of the lifestyle of the nomadic human pastoralist was the increased availability and use of psychoactive fungi. Like the agricultural activities of the ants, the behavior patterns of nomadic human societies served as an effective way for some mushrooms to expand their range.

A NEW VIEW OF HUMAN EVOLUTION

The first encounters between hominids and psilocybin-containing mushrooms may have predated the domestication of cattle in Africa by a million years or more. And during this million-year period, the mushrooms were not only gathered and eaten but probably also achieved the status of a cult. But domestication of wild cattle, a great step in human cultural evolution, by bringing humans into greater proximity to cattle, also entailed increased contact with the mushrooms, because these mushrooms grow only in the dung of cattle. As a result, the human-mushroom interspecies codependency was enhanced and deepened. It was at this time that religious ritual, calendar making, and natural magic came into their own.

Shortly after humans encountered the visionary fungi of the African grasslands, and like the leafcutter ants, we too became the dominant species of our area, and we too learned ways of "keeping the bulk of our populations safe in subterranean retreats." In our case these retreats were walled cities.

In pondering the course of human evolution, some thoughtful observers have questioned the scenario that physical anthropologists present us. Evolution in higher animals takes a long time to occur, operating in time spans of rarely less than a million years and more often in tens of millions of years. But the emergence of modern humans from the higher primates—with the enormous changes effected in brain size and behavior—transpired in fewer than three million years. Physically, in the last 100,000 years, we have apparently changed very little. But the amazing proliferation of cultures, social institutions, and linguistic systems has come so quickly that modern evolutionary biologists can scarcely account for it. Most do not even attempt an explanation.

Indeed, the absence of a theoretical model is not surprising; there is much that we do not know about the complex situation prevailing among the hominids just prior to and during the time when modern human beings were emerging onto the scene. Biological and fossil evidence clearly indicates that man is descended from primate ancestors not radically different from primate species still extant, and yet *Homo sapiens* obviously is in a class apart from other members of the order.

Thinking about human evolution ultimately means thinking about the evolution of human consciousness. What, then, are the origins of the human mind? In their explanations, some investigators have adopted a primarily cultural emphasis. They point to our unique linguistic and symbolical capabilities, our use of tools, and our ability to store information epigenetically as songs, art, books, computers, thereby creating not only culture, but also history. Others, taking a somewhat more biological approach, have emphasized our physiological and neurological peculiarities, including the exceptionally large size and complexity of the human neocortex, a great proportion of which is devoted to complex linguistic processing, storage, and retrieval of information, as well as being associated with motor systems governing activities like speech and writing. More recently the feedback interactions between cultural influence and biological ontogeny have been recognized and seen to be involved in certain human developmental oddities, such as prolonged childhood and adolescence, the delayed onset of sexual maturity, and the persistence of many essentially neonatal characteristics through adult life. Unfortunately the union of these points of view has not yet led to the recognition of the genome-shaping power of psychoactive and physioactive dietary constituents.

By 3 million years ago, and through a combination of the processes discussed above, at least three clearly recognized species of protohominids were in place in East Africa. These were *Homo africanus*, *Homo boisei*, and *Homo robustus*. Also at that time, the omnivorous *Homo habilis*, the first true hominid, had clearly emerged from a division of species that also gave rise to two vegetarian man-apes.

The grasslands expanded slowly; early hominids moved through a mosaic of grasslands and forests. These creatures, with brains proportionately only slightly larger than chimpanzees', were already walking upright and probably carrying food and tools between patches of forest which they continued to exploit for tubers and insects. Their arms were proportionately longer than ours, and they possessed a more powerful grasping hand. The evolution to upright posture and the initial expansion into a grassland niche had occurred earlier, between 9 and 5 million years ago. Unfortunately we lack fossil evidence for this earlier transition.

The hominids likely expanded their original diet of fruit and small animal kills by including underground roots, tubers, and corms. A simple digging stick would allow access to this previously untapped food source. Modern baboons on the savannah subsist largely on grass corms during certain seasons. Chimpanzees add substantial amounts of beans to their diet when they venture onto the savannah. Both baboons and chimpanzees hunt cooperatively and prey on small animals. They do not generally use tools in hunting, however, and there is no evidence that early hominids did either. Among chimps, baboons, and hominids, hunting appears to be a male activity. Early hominids hunted both cooperatively and alone.

With *Homo habilis* began a sudden and mysterious expansion of brain size. *Homo habilis*'s brain weighed an average 770 grams (27.5 ounces), compared with 530 grams (19 ounces) for competing hominids. The next two and a quarter million years brought an unusually rapid evolution in brain size and complexity. By 750,000 to 1.1 million years ago a new hominid type, *Homo erectus*, was widespread. The brain size of this new hominid was 900 to 1100 grams (2 to 2.4 pounds). Evidence is good that *Homo erectus* used tools and possessed some sort of rudimentary culture. At Choukoutien Cave in South Africa, there is good evidence of fire use along with burnt bones indicating the cooking of meat. These are attributed to *Homo erectus*, which was the earliest hominid to leave Africa, a million or so years ago.

Older theories suggested that modern humans evolved from *Homo erectus* in different locales. Increasingly, however, modern evolutionary primatologists accept the notion that modern *Homo sapiens* also arose in Africa, some 100,000 years ago, and made a second great outward migration from there to people the entire planet. At Border Cave and the Klasies River Mouth Cave in South Africa, there is evidence of the earliest modern *Homo sapiens* living in a mixed forest and grassland environment. In one of many attempts to understand this momentous transition, Charles J. Lumsden and Edward O. Wilson wrote:

> Behavioral ecologists have gradually assembled a theory to explain why the advance to an erect posture was taken, one that accounts for many of the most distinctive biological traits of modern man. The earliest man-apes shifted out of the tropical evergreen forest into more open, seasonal habitats, where they became committed to an exclusively terrestrial existence. They constructed base camps and became dependent on a division of labor, by which some individuals, probably the females, wandered less and devoted more time to the care of the young; others, primarily or exclusively the males, dispersed widely in the search for animal prey. Bi-pedalism conferred a great advantage in open-country locomotion. It also freed the arms, permitting the ancestral man-apes to use tools and to carry dead animals and other food back to the base camp. Food sharing and related forms of reciprocity automatically followed as central processes of the social life of the man-apes. So did close, long-term sexual bonding and heightened sexuality, which were put to the service of rearing the young. Many of the most distinctive forms of human social behavior are the product of this tightly interwoven complex of adaptation.[3]

One advanced hominid type followed another in the African evolutionary laboratory, and, beginning with *Homo erectus*, representatives of each type radiated across the Eurasian landmass in the interglacial periods. During each glaciation, migration out of Africa was bottled up; new hominids were "cooked" in the African ambience of intensified forces of mutation from exotic diets and climatically induced increased natural selection.

At the end of these truly remarkable three million years in the evolution of the human species, human brain size had tripled! Lumsden and Wilson call this "perhaps the fastest advance recorded for any complex organ in the whole history of life."[4] Such a remarkable rate of evolutionary change in the primary organ of a species implies the presence of extraordinary selective pressures.

Because scientists were unable to explain this tripling of the human brain size in so short a span of evolutionary time, some of the early primate paleontologists and evolutionary theorists predicted and searched for evidence of transitional skeletons. Today the idea of a "missing link" has largely been abandoned. Bipedalism, binocular vision, the opposable thumb, the throwing arm—all have been put forth as the key ingredient in the mix that caused self-reflecting humans to crystallize out of the caldron of competing hominid types and strategies. Yet all we really know is that the shift in brain size was accompanied by remarkable changes in the social organization of the hominids. They became users of tools, fire, and language. They began the process as higher animals and emerged from it 100,000 years ago as conscious, self-aware individuals.

THE REAL MISSING LINK

My contention is that mutation-causing, psychoactive chemical compounds in the early human diet directly influenced the rapid reorganization of the brain's information-processing capacities. Alkaloids in plants, specifically the hallucinogenic compounds such as psilocybin, dimethyltryptamine (DMT), and harmaline, could be the chemical factors in the protohuman diet that catalyzed the emergence of human self-reflection. The action of hallucinogens present in many common plants enhanced our information-processing activity, or environmental sensitivity, and thus contributed to the sudden expansion of the human brain size. At a later stage in this same process, hallucinogens acted as catalysts in the development of imagination, fueling the creation of internal stratagems and hopes that may well have synergized the emergence of language and religion.

In research done in the late 1960s, Roland Fischer gave small amounts of psilocybin to graduate students and then measured their ability to detect the moment when previously parallel lines became skewed. He found that performance ability on this particular task was actually improved after small doses of psilocybin.[5]

When I discussed these findings with Fischer, he smiled after explaining his conclusions, then summed up, "You see what is conclusively proven here is that under certain circumstances one is actually better informed concerning the real world if one has taken a drug than if one has not." His facetious remark stuck with me, first as an academic anecdote, later as an effort on his part to communicate something profound. What would be the consequences for evolutionary theory of admitting that some chemical habits confer adaptive advantage and thereby become deeply scripted in the behavior and even genome of some individuals?

THREE BIG STEPS FOR THE HUMAN RACE

In trying to answer that question I have constructed a scenario, some may call it fantasy; it is the world as seen from the vantage point of a mind for which the millennia are but seasons, a vision that years of musing on these matters has moved me toward. Let us imagine for a moment that we stand outside the surging gene swarm that is biological history, and that we can see the interwoven consequences of changes in diet and climate, which must certainly have been too slow to be felt by our ancestors. The scenario that unfolds involves the interconnected and mutually reinforcing effects of psilocybin taken at three different levels. Unique in its properties, psilocybin is the only substance, I believe, that could yield this scenario.

At the first, low, level of usage is the effect that Fischer noted: small amounts of psilocybin, consumed with no awareness of its psychoactivity while in the general act of browsing for food, and perhaps later consumed consciously, impart a noticeable increase in visual acuity, especially edge detection. As visual acuity is at a premium among hunter-gatherers, the discovery of the equivalent of "chemical binoculars" could not fail to have an impact on the hunting and gathering success of those individuals who availed themselves of this advantage. Partnership groups containing individuals with improved eyesight will be more successful at feeding their offspring. Because of the increase in available food, the offspring within such groups will have a higher probability of themselves reaching reproductive age. In such a situation, the outbreeding (or decline) of non-psilocybin-using groups would be a natural consequence.

Because psilocybin is a stimulant of the central nervous system, when taken in slightly larger doses, it tends to trigger restlessness and sexual arousal. Thus, at this second level of usage, by increasing instances of copulation, the mushrooms directly favored human reproduction. The tendency to regulate and schedule sexual activity within the group, by linking it to a lunar cycle of mushroom availability, may have been important as a first step toward ritual and religion. Certainly at the third and highest level of usage, religious concerns would be at the forefront of the tribe's consciousness, simply because of the power and strangeness of the experience itself.

This third level, then, is the level of the full-blown shamanic ecstasy. The psilocybin intoxication is a rapture whose breadth and depth is the despair of prose. It is wholly Other and no less mysterious to us than it was to our mushroom-munching ancestors. The boundary-dissolving qualities of shamanic ecstasy predispose hallucinogen-using tribal groups to community bonding and to group sexual activities, which promote gene mixing, higher birth rates, and a communal sense of responsibility for the group offspring.

At whatever dose the mushroom was used, it possessed the magical property of conferring adaptive advantages upon its archaic users and their group. Increased visual acuity, sexual arousal, and access to the transcendent Other led to success in obtaining food, sexual prowess and stamina, abundance of offspring, and access to realms of supernatural power. All of these advantages can be easily self-regulated through manipulation of dosage and frequency of ingestion. Chapter 4 will detail psilocybin's remarkable property of stimulating the language-forming capacity of the brain. Its power is so extraordinary that psilocybin can be considered the catalyst to the human development of language.

STEERING CLEAR OF LAMARCK

An objection to these ideas inevitably arises and should be dealt with. This scenario of human emergence may seem to smack of Lamarckism, which theorizes that characteristics acquired by an organism during its lifetime can be passed on to its progeny. The classic example is the claim that giraffes have long necks because they stretch their necks to reach high branches. This straightforward and rather common-sense idea is absolutely anathema among neo-Darwinians, who currently hold the high ground in evolutionary theory. Their position is that mutations are entirely random and that only after the mutations are expressed as the traits of organisms does natural selection mindlessly and dispassionately fulfill its function of preserving those individuals upon whom an adaptive advantage had been conferred.

Their objection can be put like this: While the mushrooms may have given us better eyesight, sex, and language when eaten, how did these enhancements get into the human genome and become innately human? Nongenetic enhancements of an organism's functioning made by outside agents retard the corresponding genetic reservoirs of those facilities by rendering them superfluous. In other words, if a necessary metabolite is common in available food, there will not be pressure to develop a trait for endogenous expression of the metabolite. Mushroom use would thus create individuals with less visual acuity, language facility, and consciousness. Nature would not provide those enhancements through organic evolution because the metabolic investment required to sustain them wouldn't pay off, relative to the tiny metabolic investment required to eat mushrooms. And yet today we all have these enhancements, without taking mushrooms. So how did the mushroom modifications get into the genome?

The short answer to this objection, one that requires no defense of Lamarck's ideas, is that the presence of psilocybin in the hominid diet changed the parameters of the process of natural selection by changing the behavioral patterns upon which that selection was operating. Experimentation with many types of foods was causing a general increase in the numbers of random mutations being offered up to the process of natural selection, while the augmentation of visual acuity, language use, and ritual activity through the use of psilocybin represented new behaviors. One of these new behaviors, language use, previously only a marginally important trait, was suddenly very useful in the context of new hunting and gathering lifestyles. Hence psilocybin inclusion in the diet shifted the parameters of human behavior in favor of patterns of activity that promoted increased language; acquisition of language led to more vocabulary and an expanded memory capacity. The psilocybin-using individuals evolved epigenetic rules or cultural forms that enabled them to survive and reproduce better than other individuals. Eventually the more successful epigenetically based styles of behavior spread through the populations along with the genes that reinforce them. In this fashion the population would evolve genetically and culturally.

As for visual acuity, perhaps the widespread need for corrective lenses among modern humans is a legacy of the long period of "artificial" enhancement of vision through psilocybin use. After all, atrophy of the olfactory abilities of human beings is thought by one school to be a result of a need for hungry omnivores to tolerate strong smells and tastes, perhaps even carrion. Trade-offs of this sort are common in evolution. The suppression of keenness of taste and smell would allow inclusion of foods in the diet that might otherwise be passed over as "too strong." Or it may indicate something more profound about our evolutionary relationship to diet. My brother Dennis has written:

> The apparent atrophy of the human olfactory system may actually represent a functional shift in a set of primitive, externally directed chemo-receptors to an interiorized regulatory function. This function may be related to the control of the human pheromonal system, which is largely under the control of the pineal gland, and which mediates, on a subliminal level, a host of psycho-sexual and psycho-social interactions between individuals. The pineal tends to suppress gonadal development and the onset of puberty, among other functions, and this mechanism may play a role in the persistence of neonatal characteristics in the human species. Delayed maturation and prolonged childhood and adolescence play a critical role in the neurological and psychological development of the individual, since they provide the circumstances which permit the post-natal development of the brain in the early, formative years of childhood. The symbolic, cognitive and linguistic stimuli that the brain experiences during this period are essential to its development and are the factors that make us the unique, conscious, symbol-manipulating, language-using beings that we are.

Neuroactive amines and alkaloids in the diet of early primates may have played a role in the biochemical activation of the pineal gland and the resulting adaptations.[6]

ACQUIRED TASTES

Humans are both attracted and repelled by substances whose taste skirts the edges of acceptability. Foods that are highly spiced or bitter or aromatic arouse strong reactions from us. We say of such a food that one must "acquire a taste" for it. This is true of foods such as soft cheese or pickled eggs, but is also and even more true of drugs. To recall one's first cigarette or first shot of bourbon is to recall an organism violently rejecting the acquisition of a particular taste. Repetition of exposure seems to be the key to acquiring a taste, which suggests that the process is complex and involves both behavioral and biochemical adaptation.

What we are talking about begins to sound strangely like the process of drug addiction. Something foreign to the body is nevertheless repeatedly introduced into it by conscious decision. The body adjusts to the new chemical regimen—and then does more than adjust: it accepts the new chemical regimen as right and proper and gives off signals of alarm should the regimen be threatened. These signals may be both psychological and physiological, and will be felt whenever the new chemical environment within the body is imperiled for any reason, including a conscious decision to discontinue use of the chemical in question.

Among the vast number of chemicals that constitute nature's molecular storehouse, we have been discussing a relatively small number of compounds that interact with the senses and the neurological processing of sensory data. These compounds include all of the psychoactive amines, alkaloids, pheromones, and hallucinogens—indeed, all compounds that can interact with any of the senses ranging from taste and smell to vision and hearing and combinations of all of these. The acquisition of a taste for these compounds, the acquisition of a behaviorally and physiologically reinforced habit, is what defines the basic chemical addiction syndrome.

These compounds have the remarkable ability to remind us of

both our frailty and our capacity for the magnificent. Drugs, like reality, seem destined to confound those who seek clear boundaries and an easy division of the world into black and white. How we meet the challenge of defining our future relationships to these compounds and to the dimensions of risk and opportunity they offer may say the final word about our potential for survival and evolution as a conscious species.

3

THE SEARCH FOR THE ORIGINAL TREE OF KNOWLEDGE

He had left the confusing flickering of the group fire and walked a few steps away to make water. The sound of his own voice came low and in the throat. *Nee nee nee nee neeeh*. She Who Feeds Us seemed unusually powerful this harvest moon night. Enchanted by the landscape transformed by intoxication and moonlight, he walked farther away from the noises of the domestic scene.

The *hekuli* was near, he could feel it. At this thought the hair on the back of his neck rose up. There was a sound like the shaking of seeds in a gourd. Then he saw the *hekuli*; it looked like an iridescent flower, mouth, or sphincter hanging in space. And there were others behind it, spinning slowly in the darkness, some one way, some another. They approached him like a school of curious jellyfish. There was a soft liquid explosion as the nearest one reached him and passed through his body. At that moment the interior of his head flared with sunrise-pink light and he was infused with the presence of the thing. Impressions followed one another too swiftly to comprehend. Time fell away, superfluids of frozen agate seemed to rush through enormous spillways. He had a sense of flinging himself happily into death, a kind of wild orgasmic paroxysm of self-affirmation. A previously inarticulate bubble of emotive intent came to his lips. Tears were running down his cheeks. He had said the words before. But he had never said and understood them in this way before. *Ta vodos! Ta vodos! I am! I am!*

HALLUCINOGENS AS THE REAL MISSING LINK

The notion we are exploring in this book is that a particular family of active chemical compounds, the indole hallucinogens, played a decisive role in the emergence of our essential humanness, of the human characteristic of self-reflection. It is important, therefore, to know just what these compounds are and the roles that they perform in nature. The defining characteristic of these hallucinogens is structural: all have a five-sided pyrrole group in association with the better known benzene ring (see Figure 28 on page 290). These molecular rings make the indoles highly reactive chemically and hence ideal molecules for metabolic activity in the high-energy world of organic life.

Hallucinogens may be psychoactive and/or physiologically active and may target many systems within the body. Some indoles are endogenous to the human body, serotonin being a good example. Many more are exogenous, found in nature and the plants we can eat. Some behave like hormones and regulate growth or rate of sexual maturation. Others influence mood and state of alertness.

The indole families of compounds that are strong visionary hallucinogens and also occur in plants are four in number:

1. *The LSD-type compounds.* Found in several related genera of morning glories and ergot, the LSD hallucinogens are rare in nature. That they are the best known of the hallucinogens is undoubtedly due to the fact that millions of doses of LSD were manufactured and sold during the 1960s. LSD is a psychedelic, but rather large doses are necessary to elicit the hallucinogenic *paradis artificiel*, of vivid and utterly transmundane hallucinations, that is produced by DMT and psilocybin at quite traditional doses. Nevertheless, many researchers have stressed the importance of the nonhallucinogenic effects of LSD and other psychedelics. These other effects in-

clude a sense of mind expansion and increased speed of thought; the ability to understand and to relate to complex issues of behavior, life patterning, and complex, decision-making networks of connective linkage.

LSD continues to be manufactured and sold in larger amounts than any other hallucinogen. It has been shown to aid in psychotherapy and the treatment of chronic alcoholism: "Wherever it has been tried, all over the world, it has proved to be an interesting treatment for a very old disease. No other drug so far has been able to match its record in salvaging tormented lives from the alcoholic scrap heap, directly, as a treatment, or indirectly, as a means of yielding valuable information."[1] Yet, as a consequence of media hysteria its potential may never be known.

2. *The tryptamine hallucinogens, especially DMT, psilocin, and psilocybin.* Tryptamine hallucinogens are found throughout the higher plant families, for example, in legumes, and psilocin and psilocybin occur in mushrooms. DMT also occurs endogenously in the human brain. For this reason, perhaps DMT should not be thought of as a drug at all, but DMT intoxication is the most profound and visually spectacular of the visionary hallucinogens, remarkable for its brevity, intensity, and nontoxicity.
3. *The beta-carbolines.* Beta-carbolines, such as harmine and harmaline, can be hallucinogenic at close to toxic levels. They are important for visionary shamanism because they can inhibit enzyme systems in the body that would otherwise depotentiate hallucinogens of the DMT type. Hence beta-carbolines can be used in conjunction with DMT to prolong and intensify visual hallucinations. This combination is the basis of the hallucinogenic brew *ayahuasca* or *yagé*, in use in Amazonian South America. Beta-carbolines are legal and until very recently were virtually unknown to the general public.
4. *The ibogaine family of substances.* These substances occur in two related African and South American tree genera, *Tabernanthe* and *Tabernamontana*. *Taber-*

nanthe iboga is a small, yellow flowered bush which has a history of usage as a hallucinogen in tropical West Africa. Its active compounds bear a structural relationship to beta-carbolines. Ibogaine is known more as a powerful aphrodisiac than as a hallucinogen. Nevertheless, in sufficient doses it is capable of inducing a powerful visionary and emotional experience.

These few numbered paragraphs above may contain the most important and exciting information that human beings have gathered concerning the natural world since the long-forgotten birth of science. More precious than the news of the anti-neutrino, more full of hope for humanity than the detection of new quasars, is the knowledge that certain plants, certain compounds, unlock forgotten doorways onto worlds of immediate experience that confound our science, and indeed, confound us. Properly understood and applied, this information can become a compass leading us back to the lost garden world of our origins.

SEEKING THE TREE OF KNOWLEDGE

In attempting to understand which indole hallucinogens and which plants might have been causally implicated in the emergence of consciousness, several important points need to be kept in mind:

The plant we are seeking must be African, since the evidence is overwhelming that the human type emerged in Africa. More specifically, the African plant should be native to grassland, as this is where our newly omnivorous ancestors learned to adapt, coordinate their bipedalism, and refine existing methods of signaling.

The plant must require no preparation; it must be active in its natural state. To suppose otherwise is to strain credulity—mixtures, compounded drugs, extracts, and concentrations all belong to later stages of culture, when human consciousness and the use of language are well established.

The plant must be continuously available to a nomadic population, easily noticed, and plentiful.

The plant must confer immediate and tangible benefits upon

individuals who are eating it. Only in that way would the plant establish and maintain itself as a part of the hominid diet.

These requirements dramatically reduce the number of contenders. Africa has a scarcity of hallucinogenic plants. This scarcity and the contrasting overabundance of such plants in the tropical New World have never been satisfactorily explained. Can it be mere coincidence that the longer an environment has been exposed to human beings, the fewer its native hallucinogens, and the fewer the species of plants generally occurring within it? Today Africa supports almost no native plants that are really good candidates for catalysis of consciousness among the evolving hominids.

Grasslands have far fewer plant species than forests. Because of this scarcity, it is highly likely that a hominid would test any grassland plant encountered for its food potential. The eminent geographer Carl Saur felt that there was no such thing as a natural grassland. He suggested that all grasslands were human artifacts, resulting from the cumulative impact of seasonal burning. He based this argument on the fact that all grassland species can be found in the understory of the forests at the edges of the grasslands whereas a very high percentage of the forest species are absent from the grasslands. Saur concluded that the grasslands are so recent that they must be seen as concomitant with the rise of fire-using human populations.[2]

WEEDING OUT THE CANDIDATES

Today only the Bwiti religion among the Fang of Gabon and Zaire can be called a truly African hallucinogenic plant cult. Conceivably, the plant that is used, *Tabernanthe iboga*, could have had some influence on prehistoric people. However, there is absolutely no evidence of its use before the early nineteenth century. At no point, for example, was it mentioned by the Portuguese, who had a long history of trade and exploration in West Africa. This lack of evidence is difficult to explain if one believes that use of the plant is very old.

Analyzed sociologically, Bwiti is a force not only for group cohesion but also for holding marriages together. Historically, divorce is a chronic source of group anxiety among the Fang. This is due to the fact that divorce is easily obtained but, once granted, must be followed by complicated, protracted, and potentially expensive negotiations with the family of the divorced partner concerning the return of a portion of the dowry.[3] Perhaps iboga, as well as being a hallucinogen, activates a pheromone promoting pair bonding. Its reputation for being an aphrodisiac could well be partially related to its promotion of pair bonding.

The plant itself is a medium-sized bush, not a native of the grasslands but of the tropical forest. It is rarely found growing outside of cultivation.

As a result of European contacts with tropical Africa, iboga became the first indole to come into vogue in Europe. Tonics based on the whole plant extract became extremely popular in France and Belgium after iboga was promoted to the public at the Paris Exposition of 1867. This crude extract was sold in Europe as Lambarene, a cure for everything from neurasthenia to syphilis and, above all, an aphrodisiac.

Not until 1901 was the alkaloid isolated. The initial wave of research that followed seemed promising. A cure for male impotence was eagerly anticipated. Yet ibogaine, once chemically characterized, was quickly forgotten. Though no evidence was ever offered that it was dangerous or addictive, the compound was placed in Schedule I, the most restrictive and controlled category, in the United States, making further research highly unlikely. Ibogaine remains to this day nearly unstudied in human beings.

What we know of the iboga cult we learned from the observations of field anthropologists. The root scrapings of the plant are taken in quite prodigious amounts. Among the Fang it is believed that they acquired this folkway during a centuries-long migration, in which they were for some time in the proximity of Pygmy people who taught them the spiritual power resident in Bwiti. The root bark of the *Tabernanthe iboga* plant contains the psychoactive portion of the plant. According to the Fang, many grams of this root material must be eaten in order to "open one's head." Lesser amounts are then effective for the remainder of a person's life.

While the iboga cult is very interesting, I do not think iboga was the catalyst of consciousness in evolving humans. As already mentioned, it has not been shown to have a long history of use and is not a grassland plant. In addition, at small doses it diminishes ordinary vision by facilitating afterimages, halos, and visual "streaking."

No plants containing LSD-type compounds are known to have been used in Africa. Nor are there any striking examples of plants that are rich in these compounds.

Peganum harmala, the giant Syrian rue, is rich in the beta-carboline harmine and today occurs wild across the arid portions of Mediterranean North Africa. There is, however, no record of its use in Africa as a hallucinogen, and in any case it must be concentrated and/or combined with DMT to activate its visionary potential.[4]

THE UR PLANT

We are left, then, by a process of elimination, with the tryptamine type hallucinogens—psilocybin, psilocin, and DMT. In a grassland environment these compounds might be expected to occur in either a dung-loving (coprophilic) mushroom containing psilocybin or in a grass containing DMT. But unless the DMT were extracted and concentrated, something beyond the technical reach of early human beings, these grasses could never supply sufficient amounts of DMT to provide an effective hallucinogen. By a process of elimination we are led to suspect a mushroom might have been involved.

When our remote ancestors moved out of the trees and on to the grasslands, they increasingly encountered hooved beasts who ate vegetation. These beasts became a major source of potential sustenance. Our ancestors also encountered the manure of these same wild cattle and the mushrooms that grow in it.

Several of these grassland mushrooms contain psilocybin: *Panaeolus* species and *Stropharia cubensis*, also called *Psilocybe cubensis* (see Figure 1). This latter is the familiar "magic mushroom," now grown by enthusiasts worldwide.[5]

Of these mushroom species, only *Stropharia cubensis* contains psilocybin in concentrated amounts and is free of nausea-producing compounds. It alone is pandemic—it occurs throughout the tropical regions, at least wherever cattle of the zebu *(Bos indicus)* type graze. This raises a number of questions. Does *Stropharia cubensis* occur

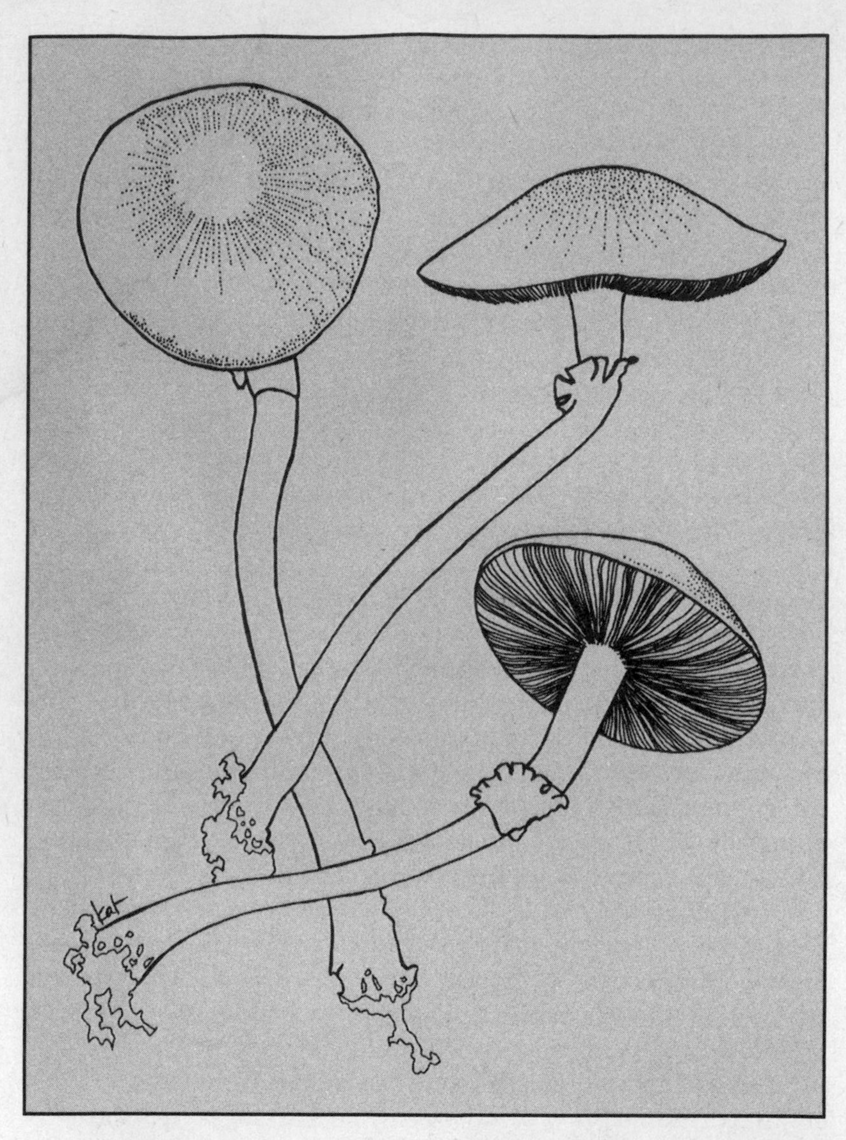

FIGURE 1. *Stropharia cubensis*. Also called *Psilocybe cubensis*. Taxonomic drawing by Kat Harrison-McKenna. From O. T. Oss and O. N. Oeric, *Psilocybin: The Magic Mushroom Grower's Guide* (Berkeley: Lux Natura Press, 1986), p. 12.

exclusively in the manure of zebu or can it occur in the manure of other cattle? How recently has it reached its various habitats? The first specimen of *Psilocybe cubensis* was collected by the American botanist Earle in Cuba in 1906, but current botanical thinking places the species' point of origin in Southeast Asia. At an archaeological dig in Thailand at a place called Non Nak Tha, which has been dated to 15,000 B.P., the bones of zebu cattle have been found coincident with human graves. *Stropharia cubensis* is common in the Non Nak Tha area today. The Non Nak Tha site suggests mushroom use was a human trait that emerged wherever human populations and cattle evolved together.

Ample evidence supports the notion that *Stropharia cubensis* is the Ur plant, our umbilicus to the feminine mind of the planet, which, when its cult, the Paleolithic cult of the Great Horned Goddess, was intact, conveyed to us such knowledge that we were able to live in a dynamic equilibrium with nature, with each other, and within ourselves. Hallucinogenic mushroom use evolved as a kind of natural habit with behavioral and evolutionary consequences. This relationship between human beings and mushrooms had to have also included cattle, the creators of the only source of the mushrooms.

The relationship is probably altogether no more than a million years old, for the era of the nomadic human hunter dates from that time. The last 100,000 years is probably a more than generous amount of time to allow for the evolution of pastoralism from its first faint glimmerings. Since the entire relationship extends no further than a million years, we are not discussing a biological symbiosis that might take many millions of years to evolve. Rather, we are talking about a deep-rooted custom, an extremely powerful natural habit.

Whatever we call the human interaction with the mushroom *Stropharia cubensis*, it has not been a static relationship, but rather a dynamic through which we have been bootstrapped to higher and higher cultural levels and levels of individual self-awareness. I believe that the use of hallucinogenic mushrooms on the grasslands of Africa gave us the model for all religions to follow. And when, after long centuries of slow forgetting, migration, and climatic change, the knowledge of the mystery was finally lost, we in our anguish traded partnership for dominance, traded harmony with nature for rape of nature, traded poetry for the sophistry of science. In short, we traded our birthright as partners in the drama of the living mind of the planet for the broken pot shards of history, warfare, neurosis, and—if we do not quickly awaken to our predicament—planetary catastrophe.

WHAT ARE PLANT HALLUCINOGENS?

In the light of their suggested importance for human evolution, it is natural to inquire what mutagens and other secondary by-products are actually doing for the plants in which they occur. This is a botanic mystery that remains controversial among evolutionary biologists even today. It has been suggested that toxic and bioactive compounds are produced in plants in order to make them unpalatable and hence undesirable as food. It has also been suggested, conversely, that such compounds were developed to attract insects or birds that pollinate or distribute seeds.

A more likely explanation for the presence of secondary compounds is based on the recognition that they are not, in fact, secondary or peripheral. Evidence for this is that alkaloids, usually regarded as secondary, are formed in the greatest amounts in tissues that are most active in overall metabolism. Alkaloids, including all the hallucinogens discussed here, are not inert end products in the plants in which they occur, but are in a dynamic state, fluctuating in both concentration and in their rate of metabolic decay. The role of these alkaloids in the chemistry of metabolism makes it clear that they are essential to the life and the survival strategy of the organism, but they are acting in ways that we do not yet understand.

One possibility is that some of these compounds may be exopheromones. Exopheromones are chemical messengers that do not act among the members of a single species, but instead, act across species lines, so that an individual influences members of a different species. Some exopheromones act in ways that allow a small group of individuals to affect a community or an entire biome.

The notion of nature as an organismic and planetary whole that mediates and controls its own development through the release of chemical messages may seem somewhat radical. Our heritage from the nineteenth century is a nature all "tooth and claw," where a pitiless and mindless natural order promotes survival of those capable of ensuring their own continued existence at the expense of competitors. Competitors, in this theory, mean all the rest of nature. Yet most evolutionary biologists have long held this classical Darwinian view of nature to be incomplete. It is now generally understood that nature, far from being endless warfare among the species, is an endless dance of diplomacy. And diplomacy is largely a matter of language.

Nature appears to maximize mutual cooperation and mutual coordination of goals. To be indispensable to the organisms with which one shares an environment—that is the strategy that ensures successful breeding and continued survival. It is a strategy in which communication and sensitivity to signal processing are paramount. These are language skills.

The idea that nature might be an organism whose interconnected components act upon and communicate with one another through the release of chemical signals into the environment is only now beginning to be carefully studied. Nature, however, tends to act with a certain economy; once developed, a given evolutionary response to a problem will be applied again and again in situations where it is appropriate.

THE TRANSCENDENT OTHER

If hallucinogens function as interspecies chemical messengers, then the dynamic of the close relationship between primate and hallucinogenic plant is one of information transfer from one species to the other. Where plant hallucinogens do not occur, such transfers of information take place with great slowness, but in the presence of hallucinogens a culture is quickly introduced to ever more novel information, sensory input, and behavior and thus is bootstrapped to higher and higher states of self-reflection. I call this the encounter with the Transcendent Other, but this is only a label, not an explanation.

From one point of view the Transcendent Other is nature correctly perceived to be alive and intelligent. From another it is the awesomely unfamiliar union of all the senses with memory of the past and anticipation of the future. The Transcendent Other is what one encounters on powerful hallucinogens. It is the crucible of the Mystery of our being, both as a species and as individuals. The Transcendent Other is Nature without her cheerfully reassuring mask of ordinary space, time, and causality.

Of course, imagining these higher states of self-reflection is not easy. For when we seek to do this we are acting as if we expect language to somehow encompass that which is, at present, beyond language, or trans-linguistic. Psilocybin, the hallucinogen unique to mushrooms, is an effective tool in this situation. Psilocybin's main synergistic effect seems ultimately to be in the domain of language. It excites vocalization; it empowers articulation; it transmutes language into something that is visibly beheld. It could have had an impact on the sudden emergence of consciousness and language use in early humans. We literally may have eaten our way to higher consciousness. In this context it is important to note that the most powerful mutagens in the natural environment occur in molds and fungi. Mushrooms and cereal grains infected by molds may have had a major influence on animal species, including primates, evolving in the grasslands.

4

PLANTS AND PRIMATES: POSTCARDS FROM THE STONED AGE

Ifi had more summers than all the fingers of his two hands. He was near now to the age when he would join the hunters at their fire. It was a great step, this short journey from the children's hut to the fire of the hunters near the song hut of the true men. It had been a long journey not through space but through time. For many years he had been pointed toward this day—the hours of practicing spear thrusts with the fire-hardened sticks that served the boys for mock weapons, Doknu's endless instruction in tracking, in reading weather signs, in remaining aware of the winds. And instruction in the magic of hunting. The boy suppressed a desire to finger the talisman that his mother had prepared for him and that now hung from his neck. He did not move. His mind seemed removed from the scene, as if viewing it from above and a little way off. He had stood thus for more than twelve hours. Unmoving, all but unwinking. "This will give you the gift of stillness. And power!" He remembered the soapy taste of the rasped root bark as he had forced it down under the watchful gaze of his teacher, Doknu. "With this you become invisible, little brother," he had said, adding in a calm voice, "Kill cleanly. Then you honor our ancestors." Ifi could feel that the moment of his truth was now nearly upon him. Under the influence of the Togna, the plant-of-power-to-sit-still, he had been brought to this desolate place and told to wait near the fresh carcass of a zebra. Doknu, his father, and his uncles had all wished him well, laughing, making promises, and using new and unfamiliar words to describe the way the village women would receive him if he succeeded. Those words had excited him for a time, but then he had settled into his wait. The Togna made this a wonderfully easy thing for the boy to do. His body seemed impervious to fatigue, and his mind drifted, delighted with scenes swimming in his head from stories and experiences told around the fire. Suddenly, and without his shifting a hair, Ifi's mind flared into total alertness: Something sounded nearby. There it was again! From the pebble-strewn wash beyond the tamarisks under which he waited came a dry sound.

Chuff. Chuff.

Ifi felt neither fear nor dread of what he was about to see. He anticipated, his muscles drew power into themselves from the shimmering air. He did not move. The lioness was enormous, and wary with the stealth of all animals in the land of the great hunters. Thinking that he was but a boulder or a tree, Ifi watched. The lioness was no more than twice his body length away. Dropping her guard, she moved forward to nuzzle the zebra's bloodied haunch. At that moment, from a center of focus hundreds of generations deep, Ifi struck—cleanly, slightly to one side of the spine and behind the shoulder blade. The scream of mingled pain and rage was earsplitting. So great was the force behind the man-boy's blow that for a moment the lioness was pinned to the ground, long enough for the boy to leap away from the claws of the dying animal. The bellies of Ifi's clan would be filled that night, and the hunter's circle would admit a new member to their boisterous and privileged ranks.

• • •

This example makes clear the way in which a beneficial plant, in this case a powerful stimulant, once discovered, can be included in the diet and thus confer an adaptive advantage. A plant can confer strength and alertness and so insure hunting success and steady food supplies. The person or group is much less threatened by certain environmental factors, which may have previously limited individuals' life spans and hence the growth of the population as a whole. Less easy to understand is the way in which plant hallucinogens might have provided similar yet different adaptive advantages. These compounds do not, for example, catalyze the immune system into higher states of activity, although this may be a secondary effect. Rather, they catalyze consciousness, that peculiar, self-reflecting ability that has reached its greatest apparent expression in human beings. They do not, however, *cause* consciousness, which is a generalized function present in some degree in all life forms. Catalysis is a speeding up of processes that are already present.

One can hardly doubt that consciousness, like the ability to resist disease, confers an immense adaptive advantage on any individual who possesses it. In the search for a causal agent capable of synergizing cognitive activity and thereby of playing a role in the emergence of the hominid, researchers might long ago have looked to plant hallucinogens were it not for our strong, almost compulsive avoidance of the idea that our exalted position in the hierarchy of nature might be somehow due to the power of plants or natural forces of any sort. Even as the nineteenth century had to come to terms with the notion of human descent from apes, we must now come to terms with the fact that those apes were stoned apes. Being stoned seems to have been our unique characteristic.

HUMAN UNIQUENESS

To seek to understand human beings is to seek to understand their uniqueness. The radical division between human beings and the rest of nature is so striking that for prescientific thinkers it was sufficient proof that we were the divinely favored portion of creation—somehow different, somehow nearer to God. After all, human beings speak, fantasize, laugh, fall in love, are capable of great acts of self-sacrifice or of cruelty; human beings create great works of art and propound theoretical and mathematical models of phenomena. And human beings distinguish themselves by the sheer numbers of kinds of substances they use and become addicted to in the environment.

HUMAN COGNITION

All the unique characteristics and preoccupations of human beings can be summed up under the heading of cognitive activities: dance, philosophy, painting, poetry, sport, meditation, erotic fantasy, politics, and ecstatic self-intoxication. We are truly *Homo sapiens*, the thinking animal; our acts are all a product of the dimension that is uniquely ours, the dimension of cognitive activity. Of thought and emotion, memory and anticipation. Of Psyche.

From observing the *ayahuasca*-using people of the Upper Amazon, it became very clear to me that shamanism is often intuitively guided group decision making. The shamans decide when the group should move or hunt or make war. Human cognition is an adaptive response that is profoundly flexible in the way it allows us to manage what in other species are genetically programmed behaviors.

We alone live in an environment that is conditioned not only by the biological and physical constraints to which all species are subject but also by symbols and language. Our human environment is conditioned by meaning. And meaning lies in the collective mind of the group.

Symbols and language allow us to act in a dimension that is "supranatural"—outside the ordinary activities of other forms of organic life. We can actualize our cultural assumptions, alter and shape the natural world in the pursuit of ideological ends and according to the internal model of the world that our symbols have empowered us to create. We do this through the elaboration of ever more effective, and hence ever more destructive, artifacts and technologies, which we feel compelled to use.

Symbols allow us to store information outside of the physical brain. This creates for us a relationship to the past very different from that of our animal companions. Finally, we must add to any analysis of the human picture the notion of self-directed modification of activity. We are able to modify our behavior patterns based on a symbolic analysis of past events, in other words, through history. Through our ability to store and recover information as images and written records, we have created a human environment as much conditioned by symbols and languages as by biological and environmental factors.

The evolutionary breakouts that led to the appearance of language and, later, writing are examples of fundamental, almost ontological, transformations of the hominid line. Besides providing us with the ability to code data outside the confines of DNA, cognitive activities allow us to transmit information across space and time. At first this amounted merely to the ability to shout a warning or a command, really little more than a modification of the cry of alarm that is a familiar feature of the behavior of social animals. Over the course of human history this impulse to communicate has motivated the elaboration of ever more effective communication techniques. But by our century, this basic ability has turned into the all-pervasive communications media, which literally engulf the space surrounding our planet. The planet swims through a self-generated ocean of messages. Telephone calls, data exchanges, and electronically transmitted entertainment create an invisible world experienced as global informational simultaneity. We think nothing of this; as a culture we take it for granted.

Our unique and feverish love of word and symbol has given us a collective gnosis, a collective understanding of ourselves and our world that has survived throughout history until very recent times. This collective gnosis lies behind the faith of earlier centuries in "universal truths" and common human values. Ideologies can be thought of as meaning-defined environments. They are invisible, yet they surround us and determine for us, though we may never realize it, what we should think about ourselves and reality. Indeed they define for us what we *can* think.

The rise of globally simultaneous electronic culture has vastly accelerated the rate at which we each can obtain information necessary to our survival. This and the sheer size of the human population as a whole have brought to a halt our physical evolution as a species. The larger a population is, the less impact mutations will have on the evolution of that species. This fact, coupled with the development of shamanism and, later, scientific medicine, has removed us from the theater of natural selection. Meanwhile libraries and electronic data bases have replaced the individual human mind as the basic hardware providing storage for the cultural data base. Symbols and languages have gradually moved us away from the style of social organization that characterized the mute nomadism of our remote ancestors and has replaced that archaic model with the vastly more complicated social organization characteristic of an electronically unified planetary society. As a result of these changes, we ourselves have become largely epigenetic, meaning that much of what we are as human beings is no longer in our genes but in our culture.

THE PREHISTORIC EMERGENCE OF HUMAN IMAGINATION

Our capacity for cognitive and linguistic activity is related to the size and organization of the human brain. Neural structures concerned with conceptualization, visualization, signification, and association are highly developed in our species. Through the act of speaking vividly, we enter into a flirtation with the domain of the imagination. The ability to associate sounds, or the small mouth noises of language, with meaningful internal images is a synesthesic activity. The most recently evolved areas of the human brain, Broca's area and the neocortex, are devoted to the control of symbol and language processing.

The conclusion universally drawn from these facts is that the highly organized neurolinguistic areas of our brain have made language and culture possible. Where the search for scenarios of human emergence and social organization is concerned, the problem is this: we know that our linguistic abilities must have evolved in response to enormous evolutionary pressures—but we do not know what these pressures were.

Where psychoactive plant use was present, hominid nervous systems over many millennia would have been flooded by hallucinogenic realms of strange and alien beauty. However, evolutionary necessity channels the organism's awareness into a narrow cul-de-sac where ordinary reality is perceived through the reducing valve of the senses. Otherwise, we would be rather poorly adapted for the rough-and-tumble of immediate existence. As creatures with animal bodies, we are aware that we are subject to a range of immediate concerns that we can ignore only at great peril. As human beings we are also aware of an interior world, beyond the needs of the animal body, but evolutionary necessity has placed that world far from ordinary consciousness.

PATTERNS AND UNDERSTANDING

Consciousness has been called awareness of awareness[1] and is characterized by novel associations and connections among the various data of experience. Consciousness is like a super nonspecific immune response. The key to the working of the immune system is the ability of one chemical to recognize, to have a key-in-lock relationship, with another. Thus both the immune system and consciousness represent systems that learn, recognize, and remember.[2]

As I write this I think of what Alfred North Whitehead said about understanding, that it is *apperception of pattern as such*. This is also a perfectly acceptable definition of consciousness. Awareness of pattern conveys the feeling that attends understanding. There presumably can be no limit to how much consciousness a species can acquire, since understanding is not a finite project with an imaginable conclusion, but rather a stance toward immediate experience. This appears self-evident from within a world view that sees consciousness as analogous to a source of light. The more powerful the light, the greater the surface area of darkness revealed. Consciousness is the moment-to-moment integration of the individual's perception of the world. How well, one could almost say how gracefully, an individual accomplishes this integration determines that individual's unique adaptive response to existence.

We are masters not only of individual cognitive activity, but, when acting together, of group cognitive activity as well. Cognitive activity within a group usually means the elaboration and manipulation of symbols and language. Although this occurs in many species, within the human species it is especially well developed. Our immense power to manipulate symbols and language gives us our unique position in the natural world. The power of our magic and our science arises out of our commitment to group mental activity, symbol sharing, meme replication (the spreading of ideas), and the telling of tall tales.

The idea, expressed above, that ordinary consciousness is the end product of a process of extensive compression and filtration, and that the psychedelic experience is the antithesis of this construction, was put forward by Aldous Huxley, who contrasted this with the psychedelic experience. In analyzing his experiences with mescaline, Huxley wrote:

> I find myself agreeing with the eminent Cambridge philosopher, Dr. C. D. Broad, "that we should do well to consider the suggestion that the function of the brain and nervous system and sense organs is in the main *eliminative* and not productive." The function of the brain and nervous system is to protect us from being overwhelmed and confused by this mass of largely useless and irrelevant knowledge, by shutting out most of what we should otherwise perceive or remember at any moment, and leaving only that very small and special selection which is likely to be practically useful. According to such a theory, each one of us is potentially Mind at Large. But in so far as we are animals, our business is at all costs to survive. To make biological survival possible, Mind at Large has to be funnelled through the reducing valve of the brain and nervous system. What comes out at the other end is a measly trickle of the kind of consciousness which will help us to stay alive on the surface of this particular planet. To formulate and express the contents of this reduced awareness, man has invented and endlessly elaborated those symbol-systems and implicit philosophies which we call languages. Every individual is at once the beneficiary and the victim of the linguistic tradition into which he has been born. That which, in the language of religion, is called "this world" is the universe of reduced awareness, expressed, and, as it were, petrified by language. The various "other worlds" with which human beings erratically make contact are so many elements in the totality of the awareness belonging to Mind at Large. . . . Temporary by-passes may be acquired either spontaneously, or as the result of deliberate "spiritual exercises," . . . or by means of drugs.[3]

What Huxley did not mention was that drugs, specifically the plant hallucinogens, can reliably and repeatedly open the floodgates of the reducing valve of consciousness and expose the individual to the full force of the howling Tao. The way in which we internalize the impact of this experience of the Unspeakable, whether encountered through psychedelics or other means, is to generalize and extrapolate our world view through acts of imagination. These acts of imagination represent our adaptive response to information concerning the outside world that is conveyed to us by our senses. In our species, culture-specific, situation-specific syntactic software in the form of language can compete with and sometimes replace the instinctual world of hard-wired animal behavior. This means that we can learn and communicate experience and thus put maladaptive behaviors behind us. We can collectively recognize the virtues of peace over war, or of cooperation over struggle. We can change.

As we have seen, human language may have arisen when primate organizational potential was synergized by plant hallucinogens. The psychedelic experience inspired us to true self-reflective thought in the first place and then further inspired us to communicate our thoughts about it.

Others have sensed the importance of hallucinations as catalysts of human psychic organization. Julian Jaynes's theory, presented in his controversial book *The Origin of Consciousness in the Breakdown of the Bicameral Mind*,[4] makes the point that major shifts in human self-definition may have occurred even in historical times. He proposes that through Homeric times people did not have the kind of interior psychic organization that we take for granted. Thus, what we call ego was for Homeric people a "god." When danger threatened suddenly, the god's voice was heard in the individual's mind; an intrusive and alien psychic function was expressed as a kind of metaprogram for survival called forth under moments of great stress. This psychic function was perceived by those experiencing it as the direct voice of a god, of the king, or of the king in the afterlife. Merchants and traders moving from one society to another brought the unwelcome news that the gods were saying different things in different places, and so cast early seeds of doubt. At some point people integrated this previously autonomous function, and each person *became* the god and reinterpreted the inner voice as the "self" or, as it was later called, the "ego."

Jaynes's theory has been largely dismissed. Regrettably his book on the impact of hallucinations on culture, though 467 pages in length, manages to avoid discussion of hallucinogenic plants or drugs nearly entirely. By this omission Jaynes deprived himself of a mechanism that could reliably drive the kind of transformative changes he saw taking place in the evolution of human consciousness.

CATALYZING CONSCIOUSNESS

The impact of hallucinogens in the diet has been more than psychological; hallucinogenic plants may have been the catalysts for everything about us that distinguishes us from other higher primates, for all the mental functions that we associate with humanness. Our society more than others will find this theory difficult to accept, because we have made pharmacologically obtained ecstasy a taboo. Like sexuality, altered states of consciousness are taboo because they are consciously or unconsciously sensed to be entwined with the mysteries of our origin—with where we came from and how we got to be the way we are. Such experiences dissolve boundaries and threaten the order of the reigning patriarchy and the domination of society by the unreflecting expression of ego. Yet consider how plant hallucinogens may have catalyzed the use of language, the most unique of human activities.

One has, in a hallucinogenic state, the incontrovertible impression that language possesses an objectified and visible dimension, which is ordinarily hidden from our awareness. Language, under such conditions, is seen, is beheld, just as we would ordinarily see our homes and normal surroundings. In fact our ordinary cultural environment is correctly recognized, during the experience of the altered state, as the bass drone in the ongoing linguistic business of objectifying the imagination. In other words, the collectively designed cultural environment in which we all live is the objectification of our collective linguistic intent.

Our language-forming ability may have become active through the mutagenic influence of hallucinogens working directly on organelles that are concerned with the processing and generation of signals. These neural substructures are found in various portions of the brain, such as Broca's area, that govern speech formation. In other words, opening the valve that limits consciousness forces utterance, almost as if the word is a concretion of meaning previously felt but left unarticulated. This active impulse to speak, the "going forth of the word," is sensed and described in the cosmogonies of many peoples.

Psilocybin specifically activates the areas of the brain concerned with processing signals. A common occurrence with psilocybin intoxication is spontaneous outbursts of poetry and other vocal activity such as speaking in tongues, though in a manner distinct from ordinary glossolalia. In cultures with a tradition of mushroom use, these phenomena have given rise to the notion of discourse with spirit doctors and supernatural allies. Researchers familiar with the territory agree that psilocybin has a profoundly catalytic effect on the linguistic impulse.

Once activities involving syntactic self-expression were established habits among early human beings, the continued evolution of language in environments where mushrooms were scarce or unavailable permitted a tendency toward the expression and emergence of the ego. If the ego is not regularly and repeatedly dissolved in the unbounded hyperspace of the Transcendent Other, there will always be slow drift away from the sense of self as part of nature's larger whole. The ultimate consequence of this drift is the fatal ennui that now permeates Western civilization.

The connection between mushrooms and language was brilliantly anticipated by Henry Munn in his essay "The Mushrooms of Language":

> Language is an ecstatic activity of signification. Intoxicated by the mushrooms, the fluency, the ease, the aptness of expression one becomes capable of are such that one is astounded by the words that issue forth from the contact of the intention of articulation with the matter of experience. The spontaneity the mushrooms liberate is not only perceptual, but linguistic. For the shaman, it is as if existence were uttering itself through him.[5]

THE FLESH MADE WORD

The evolutionary advantages of the use of speech are both obvious and subtle. Many unusual factors converged at the birth of human language. Obviously speech facilitates communication and cognitive activity, but it also may have had unanticipated effects on the whole human enterprise.

Some neurophysiologists have hypothesized that the vocal vibration associated with human use of language caused a kind of cleansing of the cerebrospinal fluid. It has been observed that vibrations can precipitate and concentrate small molecules in the spinal fluid, which bathes and continuously purifies the brain. Our ancestors may have, consciously or unconsciously, discovered that vocal sound cleared the chemical cobwebs out of their heads. This practice may have affected the evolution of our present-day thin skull structure and proclivity for language. A self-regulated process as simple as singing might well have positive adaptive advantages if it also made the removal of chemical waste from the brain more efficient. The following excerpt supports this provocative idea:

> Vibrations of human skull, as produced by loud vocalization, exert a massaging effect on the brain and facilitate elution of metabolic products from the brain into the cerebrospinal fluid (CSF). . . . The Neanderthals had a brain 15% larger than we have, yet they did not survive in competition with modern humans. Their brains were more polluted, because their massive skulls did not vibrate and therefore the brains were not sufficiently cleaned. In the evolution of the modern humans the thinning of cranial bones was important.[6]

As already discussed, hominids and hallucinogenic plants must have been in close association for a long span of time, especially if we want to suggest that actual physical changes in the human genome resulted from the association. The structure of the soft palate in the human infant and timing of its descent is a recent adaptation that facilitates the acquisition of language. No other primate exhibits this characteristic. This change may have been a result of selective pressure on mutations originally caused by the new omnivorous diet.

WOMEN AND LANGUAGE

Women, the gatherers in the Archaic hunter-gatherer equation, were under much greater pressure to develop language than were their male counterparts. Hunting, the prerogative of the larger male, placed a premium on strength, stealth, and stoic waiting. The hunter was able to function quite well on a very limited number of linguistic signals, as is still the case among hunting peoples such as the !Kung or the Maku.

For gatherers, the situation was different. Those women with the largest repertoire of communicable images of foods and their sources and secrets of preparation were unquestionably placed in a position of advantage. Language may well have arisen as a mysterious power possessed largely by women—women who spent much more of their waking time together—and, usually, talking—than did men, women who in all societies are seen as group-minded, in contrast to the lone male image, which is the romanticized version of the alpha male of the primate troop.

The linguistic accomplishments of women were driven by a need to remember and describe to each other a variety of locations and landmarks as well as numerous taxonomic and structural details about plants to be sought or avoided. The complex morphology of the natural world propelled the evolution of language toward modeling of the world beheld. To this day a taxonomic description of a plant is a Joycean thrill to read: "Shrub 2 to 6 feet in height, glabrous throughout. Leaves mostly opposite, some in threes or uppermost alternate, sessile, linear-lanceolate or lanceolate, acute or acuminate. Flowers solitary in axils, yellow, with aroma, pedicellate. Calyx campanulate, petals soon caducous, obovate" and so on for many lines.

The linguistic depth women attained as gatherers eventually led to a momentous discovery: the discovery of agriculture. I call it momentous because of its consequences. Women realized that they could simply grow a restricted number of plants. As a result, they learned the needs of only those few plants, embraced a sedentary lifestyle, and began to forget the rest of nature they had once known so well.

At that point the retreat from the natural world began, and the dualism of humanity versus nature was born. As we will soon see, one of the places where the old goddess culture died, Çatal Hüyük, in present-day Anatolian Turkey, is the very place where agriculture may have first arisen. At places like Çatal Hüyük and Jericho, humans and their domesticated plants and animals became for the first time physically and psychologically separate from the life of untamed nature and the howling unknown. Use of hallucinogens can only be sanctioned in hunting and gathering societies. When agriculturists use these plants, they are unable to get up at dawn the morning after and go hoe the fields. At that point, corn and grain become gods—gods that symbolize domesticity and hard labor. These replace the old goddesses of plant-induced ecstasy.

Agriculture brings with it the potential for overproduction, which leads to excess wealth, hoarding, and trade. Trade leads to cities; cities isolate their inhabitants from the natural world. Paradoxically, more efficient utilization of plant resources through agriculture led to a breaking away from the symbiotic relationship that had bound human beings to nature. I do not mean this metaphorically. The ennui of modernity is the consequence of a disrupted quasi-symbiotic relationship between ourselves and Gaian nature. Only a restoration of this relationship in some form is capable of carrying us into a full appreciation of our birthright and sense of ourselves as complete human beings.

5
HABIT AS CULTURE AND RELIGION

At regular intervals that were probably lunar, the ordinary activities of the small nomadic group of herders were put aside. Rains usually followed the new moon in the tropics, making mushrooms plentiful. Gatherings took place at night; night is the time of magical projection and hallucinations, and visions are more easily obtained in darkness. The whole clan was present from oldest to youngest. Elders, especially shamans, usually women but often men, doled out each person's dose. Each clan member stood before the group and reflectively chewed and swallowed the body of the Goddess before returning to his or her place in the circle. Bone flutes and drums wove within the chanting. Line dances with heavy foot stamping channeled the energy of the first wave of visions. Suddenly the elders signal silence.

In the motionless darkness each mind follows its own trail of sparks into the bush while some people keen softly. They feel fear, and they triumph over fear through the strength of the group. They feel relief mingled with wonder at the beauty of the visionary expanse; some spontaneously reach out to those nearby in simple affection and an impulse for closeness or in erotic desire. An individual feels no distance between himself or herself and the rest of the clan or between the clan and the world. Identity is dissolved in the higher wordless truth of ecstasy. In that world, all divisions are overcome. There is only the One Great Life; it sees itself at play, and it is glad.

. . .

The impact of plants on the evolution of culture and consciousness has not been widely explored, though a conservative form of this notion appears in R. Gordon Wasson's *The Road to Eleusis*. Wasson does not comment on the emergence of self-reflection in hominids, but does suggest hallucinogenic mushrooms as the causal agent in the appearance of spiritually aware human beings and the genesis of religion. Wasson feels that omnivorous foraging humans would have sooner or later encountered hallucinogenic mushrooms or other psychoactive plants in their environment:

> As man emerged from his brutish past, thousands of years ago, there was a stage in the evolution of his awareness when the discovery of the mushroom (or was it a higher plant?) with miraculous properties was a revelation to him, a veritable detonator to his soul, arousing in him sentiments of awe and reverence, and gentleness and love, to the highest pitch of which mankind is capable, all those sentiments and virtues that mankind has ever since regarded as the highest attribute of his kind. It made him see what this perishing mortal eye cannot see. How right the Greeks were to hedge about this Mystery, this imbibing of the potion with secrecy and surveillance! . . . Perhaps with all our modern knowledge we do not need the divine mushroom anymore. Or do we need them more than ever? Some are shocked that the key even to religion might be reduced to a mere drug. On the other hand, the drug is as mysterious as it ever was: "like the wind that comes we know not whence nor why." Out of a mere drug comes the ineffable, comes ecstasy. It is not the only instance in the history of humankind where the lowly has given birth to the divine.[1]

Scattered across the African grasslands, the mushrooms would be especially noticeable to hungry eyes because of their inviting smell and unusual form and color. Once having experienced the state of consciousness induced by the mushrooms, foraging humans would return to them repeatedly, in order to reexperience their bewitching novelty. This process would create what C. H. Waddington called a "creode,"[2] a pathway of developmental activity, what we call a habit.

ECSTASY

We have already mentioned the importance of ecstasy for shamanism. Among early humans a preference for the intoxication experience was ensured simply because the experience was ecstatic. "Ecstatic" is a word central to my argument and preeminently worthy of further attention. It is a notion that is forced on us whenever we wish to indicate an experience or a state of mind that is cosmic in scale. An ecstatic experience transcends duality; it is simultaneously terrifying, hilarious, awe-inspiring, familiar, and bizarre. It is an experience that one wishes to have over and over again.

For a minded and language-using species like ourselves, the experience of ecstasy is not perceived as simple pleasure but, rather, is incredibly intense and complex. It is tied up with the very nature of ourselves and our reality, our languages, and our imagings of ourselves. It is fitting, then, that it is enshrined at the center of shamanic approaches to existence. As Mircea Eliade pointed out, shamanism and ecstasy are at root one concern:

> This shamanic complex is very old; it is found, in whole or in part, among the Australians, the archaic peoples of North and South America, in the polar regions, etc. The essential and defining element of shamanism is ecstasy—the shaman is a specialist in the sacred, able to abandon his body and undertake cosmic journeys "in the spirit" (in trance). "Possession" by spirits, although documented in a great many shamanisms, does not seem to have been a primary and essential element. Rather, it suggests a phenomenon of degeneration; for the supreme goal of the shaman is to abandon his body and rise to heaven or descend

into hell—not to let himself be "possessed" by his assisting spirits, by demons or the souls of the dead; the shaman's ideal is to master these spirits, not to let himself be "occupied" by them.[3]

Gordon Wasson added these observations on ecstasy:

In his trance the shaman goes on a far journey—the place of the departed ancestors, or the nether world, or there where the gods dwell—and this wonderland is, I submit, precisely where the hallucinogens take us. They are a gateway to ecstasy. Ecstasy in itself is neither pleasant nor unpleasant. The bliss or panic into which it plunges you is incidental to ecstasy. When you are in a state of ecstasy, your very soul seems scooped out from your body and away it goes. Who controls its flight: Is it you, or your "subconscious," or a "higher power"? Perhaps it is pitch dark, yet you see and hear more clearly than you have ever seen or heard before. You are at last face to face with Ultimate Truth: this is the overwhelming impression (or illusion) that grips you. You may visit Hell, or the Elysian fields of Asphodel, or the Gobi desert, or Arctic wastes. You know awe, you know bliss, and fear, even terror. Everyone experiences ecstasy in his own way, and never twice in the same way. Ecstasy is the very essence of shamanism. The neophyte from the great world associates the mushrooms primarily with visions, but for those who know the Indian language of the shaman the mushrooms "speak" through the shaman. The mushroom is the Word: *es habla*, as Aurelio told me. The mushroom bestows on the *curandero* what the Greeks called *Logos*, the Aryan *Vac*, Vedic *Kavya*, "poetic potency," as Louis Renous put it. The divine afflatus of poetry is the gift of the entheogen. The textual exegete skilled only in dissecting the cruces of the verses lying before him is of course indispensable and his shrewd observations should have our full attention, but unless gifted with *Kavya*, he does well to be cautious in discussing the higher reaches of Poetry. He dissects the verses but knows not ecstasy, which is the soul of the verses.[4]

In claiming that religion originated when hominids encountered hallucinogenic alkaloids, Wasson was at odds with Mircea Eliade. Eliade considered what he called "narcotic" shamanism to be decadent. He felt that if individuals cannot achieve ecstasy without drugs, then their culture is probably in a decadent phase. The use of the word "narcotic"—a term usually reserved for soporifics—to describe this form of shamanism betrays a botanical and pharmacological naïveté. Wasson's view, which I share, is precisely the opposite: the presence of a hallucinogen indicates that shamanism is authentic and alive; the late, decadent phase of shamanism is characterized by elaborate rituals, ordeals and reliance on pathological personalities. Where these phenomena are central, shamanism is well on its way to becoming simply "religion."[5]

And at its fullest, shamanism is not simply religion, it is a dynamic connection into the totality of life on the planet. If, as suggested earlier, hallucinogens operate in the natural environment as message-bearing molecules, exopheromones, then the relationship between primate and hallucinogenic plant signifies a transfer of information from one species to another. The benefits to the mushroom arise out of the hominid domestication of cattle and hence the expansion of the niche occupied by the mushroom. Where plant hallucinogens do not occur, cultural innovation occurs very slowly, if at all, but we have seen that in the presence of hallucinogens a culture is regularly introduced to ever more novel information, sensory input, and behavior and thus is moved to higher and higher states of self-reflection. The shamans are the vanguard of this creative advance.

How, specifically, might the consciousness-catalyzing properties of plants have played a role in the emergence of culture and religion? What *was* the effect of this folkway, this promotion of language-using, thinking, but stoned hominids into the natural order? I believe that the natural psychedelic compounds acted as feminizing agents that tempered and civilized the egocentric values of the solitary hunter-individual with the feminine concerns for child-rearing and group survival. The prolonged and repeated exposure to the psychedelic experience, the Wholly Other rupture of the mundane plane caused by the hallucinogenic ritual ecstasy, acted steadily to dissolve that portion of the psyche which we moderns call the ego. Wherever and whenever the ego function began to form, it was akin to a calcareous tumor or a blockage in the energy of the psyche. The use of psychedelic plants in a context of shamanic initiation dissolved—as it dissolves today—the knotted structure of the ego into undifferentiated feeling, what Eastern philosophy calls the Tao. This dissolving of individual identity into the Tao is the goal of much of Eastern thought and has traditionally been recognized as the key to psychological health and balance for both the group and the individual. To appraise our dilemma correctly, we need to appraise what this loss of Tao, this loss of collective connection to the Earth, has meant for our humanness.

MONOTHEISM

We in the West are the inheritors of a very different understanding of the world. Loss of connection to the Tao has meant that the psychological development of Western civilization has been markedly different from the East's. In the West there has been a steady focus on the ego and on the god of the ego—the monotheistic ideal. Monotheism exhibits what is essentially a pathological personality pattern projected onto the ideal of God: the pattern of the paranoid, possessive, power-obsessed male ego. This God is not someone you would care to invite to a garden party. Also interesting is that the Western ideal is the only formulation of deity that has no relationship with woman at any point in the theological myth. In ancient Babylon Anu was paired with his consort Inanna; Grecian religion assigned Zeus a wife, many consorts, and daughters. These heavenly pairings are typical. Only the god of Western civilization has no mother, no sister, no female consort, and no daughter.

Hinduism and Buddhism have maintained traditions of techniques of ecstasy that include, as stated in the *Yogic Sutras of Patanjali*, "light filled herbs," and the rituals of these great religions give ample scope for the expression and appreciation of the feminine. Sadly, the Western tradition has suffered a long, sustained break with the sociosymbiotic relationship to the feminine and the mysteries of organic life that can be realized through shamanic use of hallucinogenic plants.

Modern religion in the West is a set of social patterns, or a set of anxieties centered on a particular moral structure and view of obligation. Modern religion is rarely an experience of setting aside the ego. Since the 1960s, the spread of popular cults of trance and dance, such as disco and reggae, is an inevitable and healthy counter to the generally moribund form religious expression has taken on in Western and high-tech culture. The connection between rock and roll and psychedelics is a shamanic connection; trance, dance, and intoxication make up the Archaic formula for both religious celebration and a guaranteed good time.

The global triumph of Western values means we, as a species, have wandered into a state of prolonged neurosis because of the absence of a connection to the unconscious. Gaining access to the unconscious through plant hallucinogen use reaffirms our original bond to the living planet. Our estrangement from nature and the unconscious became entrenched roughly two thousand years ago, during the shift from the Age of the Great God Pan to that of Pisces that occurred with the suppression of the pagan mysteries and the rise of Christianity. The psychological shift that ensued left European civilization staring into two millennia of religious mania and persecution, warfare, materialism, and rationalism.

The monstrous forces of scientific industrialism and global politics that have been born into modern times were conceived at the time of the shattering of the symbiotic relationships with the plants that had bound us to nature from our dim beginnings. This left each human being frightened, guilt-burdened, and alone. Existential man was born.

Terror of being was the placenta that accompanied the birth of Christianity, the ultimate cult of domination by the unconstrained male ego. The abandonment of the ego-dissolving rites of the visionary plants had allowed what began as an individually maladaptive style to become the guiding image of the entire social organism. From within the context of an unchecked growth of dominator values and history told from a dominator point of view, we need to turn attention back toward the Archaic way of vision plants and the Goddess.

PATHOLOGICAL MONOTHEISM

The drive for unitary wholeness within the psyche, which is to a degree instinctual, can nevertheless become pathological if pursued in a context in which dissolution of boundaries and rediscovery of the ground of being has been made impossible. Monotheism became the carrier of the dominator model, the Apollonian model of the self as solar and complete in its masculine expression. As a result of this pathological model, the worth and power of emotion and the natural world have been devalued and replaced by a narcissistic fascination with the abstract and the metaphysical. This attitude has proved a double-edged sword, it has given science explanatory power and its capacity for moral bankruptcy.

Dominator culture has shown a remarkable ability to redesign itself to meet changing levels of technology and collective self-awareness. In all its manifestations, monotheism has been and remains the single most stubborn force resisting perception of the primacy of the natural world. Monotheism strenuously denies the need to return to a cultural style that periodically places the ego and its values in perspective through contact with a boundary-dissolving immersion in the Archaic mystery of plant-induced, hence mother-associated, psychedelic ecstasy and wholeness, what Joyce called the "mama matrix most mysterious."

ARCHAIC SEXUALITY

This is not to imply that the life of the nomadic pastoralist is free of anxiety. Doubtless jealousy and possessiveness persisted among mushroom-using archaic humans, if only as a vestige of hierarchical organization in the social forms of protohominids. Observation of modern primates—of their dominance games and their violently enforced hierarchical structure—suggests that protohominid societies that were premushroom may well have been dominator in style. Thus, we may have experienced no more than a brief abandonment of the dominator style—a brief tendency toward a true dynamic and conscious equilibrium with nature, at variance with our primate past and too soon crushed beneath the chariot wheels of historical process. Since the abandonment of our sojourn with mushroom use in the African Eden, we have only become progressively more bestial in our treatment of one another.

An open and nonproprietary approach to sexuality is fundamental to the partnership model. But this tendency was synergized and strengthened by the orgiastic behavior that was certainly a part of the African Goddess/mushroom religion. Group sexual activity within a small tribe of hunter-gatherers and group experiences with hallucinogens acted to dissolve boundaries and differences between people and to promote the open and unstructured sexuality that is naturally a part of nomadic tribalism. (This is not to imply that contemporary mushroom rituals are "orgies," despite what a small sensation-hungry segment of the public may choose to believe.)

IBOGAINE AMONG THE FANG

The Bwiti cults of West Africa, discussed in Chapter 3, offer an instructive example: use of a hallucinogenic indole-containing plant provides not only visionary ecstasy but also what its users call "open heartedness." This quality, a caring awareness of others, is widely believed to explain the internal cohesiveness of Fang society and the ability of Bwitists among the Fang to resist commercial and missionary incursions into their cultural integrity:

> Neither Bwitists nor Fang felt they could eradicate ritual sin or evil in the world. This incapacity means that men have to celebrate. Good and bad walk together. As Fang frequently enough told missionaries, "We have two hearts, good and bad." Early missionaries, aware of these self-confessed contradictions, evangelized with the promise of "one heartedness" in Christianity. But Fang by and large did not find it there. For many, Christian one heartedness was a constriction of their selves. While "one heartedness" is celebrated in Bwiti, it is a one heartedness which is coagulated out of a flow of many qualities from one state to another. It is goodness achieved in the presence of badness, an aboveness achieved in the presence of belowness. It is an emergent quality energized in the presence of its opposite.[6]

Paradoxically ibogaine, the indole hallucinogen responsible for the pharmacological activity of the Bwiti plant (*Tabernanthe iboga*), is widely recognized both as a factor holding married couples together in the face of Fang institutions like easy divorce and as an aphrodisiac. It is perhaps one of the few plants of the many dozens claimed to be aphrodisiacs that actually performs as advertised.[7] Most other candidates for the title are in fact merely stimulants that can cause a generalized arousal and sustained erection.

Ibogaine seems actually to change, to deepen, and to enhance the psychological mechanisms that lie behind sexual drive; one experiences a simultaneous sense of detachment and involvement that is empowering. Yet in situations where sexual activity is neither sanctioned nor appropriate, ibogaine does not cause, or even raise the possibility of, sexual behavior. In these situations it functions much as *ayahuasca* functions among its traditional users; as a boundary-dissolving visionary hallucinogen. Here is another example of research only waiting for social attitudes to change in order to be done. If the impact of ibogaine on sexual dysfunction is found to be congruent with its folklore, then further research might be especially promising.

These powerful plants that change our relationship to our sexuality, and our view of self and world, are the special province of peoples whom we are accustomed to thinking of as primitive. This is but one more indication of the extent to which unconsciously imbibed dominator attitudes have robbed us of participation in the wider and richer world of eros and the spirit.

For easily discerned reasons, the dominator societies that arose to replace partnership societies were far less eager to suppress group sexual activities than they were to suppress the hallucinogenic mushroom religion. Group sexual activity without the dissolution of the dominator ego would help the most ego-obsessed males gain power and rise in the social hierarchy. Since domination of others ultimately includes sexual domination as well, this would explain the persistence of orgies and group sexual activities in many of the mystery religions, at the festivals of Dionysus and the Roman Saturnalia, and within paganism generally long after the heart of the pagan world had ceased to beat. Eventually, however, the dominator anxiety about the establishing of clear lines of male paternity outweighed all other considerations. Then ego domination finally achieved complete preeminence. Through Christianity's ruthless extermination of all heterodoxy, orgies were recognized and suppressed as the subversive, boundary-dissolving activities that they are.

CONTRASTS IN SEXUAL POLITICS

Several important contrasts emerge from a comparison of the ego-based dominator society and the nonrigid, psychologically unbounded partnership society. Much diminished in the partnership model is the proprietary attitude of men toward women that is so centrally a part of the dominator model. Less prominent as well is the tendency for women to seek extended commitment to pair bonding from men in the pursuit of security and vicarious social ranking. Family organization is not rigid and hierarchical. Children are raised by an extended family of cousins and siblings, aunts and uncles, and former and current sexual partners of their parents. In such a milieu, a child has many different relationships and a variety of role models. Group values are not usually at odds with that of the individual or his or her mate and children. Adolescent sexual experimentation is expected and encouraged. Couples may bond for any number of reasons related to themselves and the welfare of the group; such bonding may be—but is not necessarily—lifelong. Sexuality is rarely taboo in such societies, only becoming so as a result of contact with dominator values.

In dominator society, men tend to choose sexual partners who are young, healthy, and capable of bearing many children. And the strategy of women within a dominator society is often to bond with an older man who, by being in control of group resources (food, land, or other women), could ensure that a woman's worth won't be devalued as she becomes older and passes out of her childbearing years. In the ideal partnership society, older men may have sexual relations with younger women, but without threatening the bonds that have been formed with older women; however, women are not driven to seek reproductive security under the protection of older men.

This situation arose because power did not lie exclusively with aging and powerful males. Rather, power was distributed between men and women and through all age groups. Ultimate power in such societies was the power to create and sustain life and so was naturally imaged as female—the power of the great Goddess.

Jean Baker Miller pointed out that the so-called need to control and dominate others is psychologically a function, *not* of a feeling of power, but of a feeling of powerlessness. Distinguishing between "power for oneself and power over others," she writes: "In a basic sense, the greater the development of each individual the more able, more effective, and less needy of limiting or restricting others she or he will be."[8]

Partnership societies do not simply replace a patriarchy with a matriarchy; such concepts are too limited and gender bound. The real difference here is between a society based on partnership and roles appropriate to age, size, and level of skill and a society in which a dominance hierarchy is maintained at the expense of the full expression and social utilization of the individuals within the group. In the partnership situation the lack of concepts based on property and ego inflation made jealousy and possessiveness less of a problem.

The generally hostile attitude of dominator society toward sexual expression can be traced to the terror that the dominator ego feels in any situation in which boundaries are dissolved, even the most pleasurable and natural of situations. The French notion of orgasm as *petit mort* perfectly encapsulates the fear and fascination that boundary-dissolving orgasm holds for dominator cultures.

6

THE HIGH PLAINS OF EDEN

Angi and her sister, along with some of the other girls of the sib, crowded around the doorway into the chapel. The cowhide covering that usually hid the interior from view had been removed. This was the time of the fall festival celebrating the bounty of the Great Goddess. The great women of the city, their hair greased back, their breasts and thighs covered with the gray-blue color of ceremonial ash, were kneeling and singing around the festooned, entranced figure of the Goddess. She was resplendent, lying on the horn recliner with bunches of flowers and offerings of pine nuts heaped about her. Watching through the flickering of many lamps, the young observers never so much as imagined that what they saw was not the Goddess herself, her pregnant form rising and falling in deep sleep, but a wooden statue inset with the fine obsidian for which the city was famous and rubbed with generations of pigments and fat until her skin shone with the same deep ebony luster as that of the city's people.

In a small open space at the foot of the Goddess, three of the shamans of the highest, most secret order danced slowly in vulture costumes, whose shadows mingled hypnotically with similar vultures painted on the whitewashed walls. At the conclusion of the dance, richly painted wooden vessels with lids were brought from a niche in the wall and unwrapped from covers of dyed and woven fabric. Every one present, even our little spies at the doorway, knew that the mushroom, She of Many Names, was within. And the sacrament was brought forth and distributed to be eaten among the women present. It was a rare privilege for the girls to be ignored and so allowed to witness the Harvest Mother mysteries—a mark of their rising status with the women, really. Each knew that in a few years she would take her place as an initiate into the ritual which they now beheld but could not understand. Though she was only eight, and her sister Slinga was six, Angi knew that what they saw, no man of the city had ever seen. The men's mysteries were different, also secret, also never spoken of.

THE TASSILI PLATEAU

Archaeological evidence for these speculative ideas can be found in the Sahara Desert of southern Algeria in an area called the Tassili-n-Ajjer Plateau. A curious geological formation, the plateau is like a labyrinth, a vast badlands of stone escarpments that have been cut by the wind into many perpendicular narrow corridors. Aerial photographs give the eerie impression of an abandoned city (Figure 2).

In the Tassili-n-Ajjer, rock paintings date from the late Neolithic to as recently as two thousand years ago. Here are the earliest known depictions of shamans with large numbers of grazing cattle. The shamans are dancing with fists full of mushrooms and also have mushrooms sprouting out of their bodies (Figure 3). In one instance they are shown running joyfully, surrounded by the geometric structures of their hallucinations[1] (Figure 4). The pictorial evidence seems incontrovertible.

Images similar to those of the Tassili occur in pre-Columbian Peruvian textiles. In these textiles the shamans hold objects that may be mushrooms but may also be chopping tools. With the Tassili frescoes, however, the case is clear. At Matalen-Amazar and Ti-n-Tazarift on the Tassili, the dancing shamans clearly have mushrooms in their hands and sprouting from their bodies.

The pastoral peoples who produced the Tassili paintings gradually moved out of Africa over a long period of time, from twenty thou-

FIGURE 2. Aerial photo of the region of Tamrit, Ti-n-Bedjadj of the Tassili-n-Ajjer Plateau. From *The Search for the Tassili Frescoes* by Henri Lhote (New York: E. P. Dutton, 1959), Figure 71, pp. 184–185.

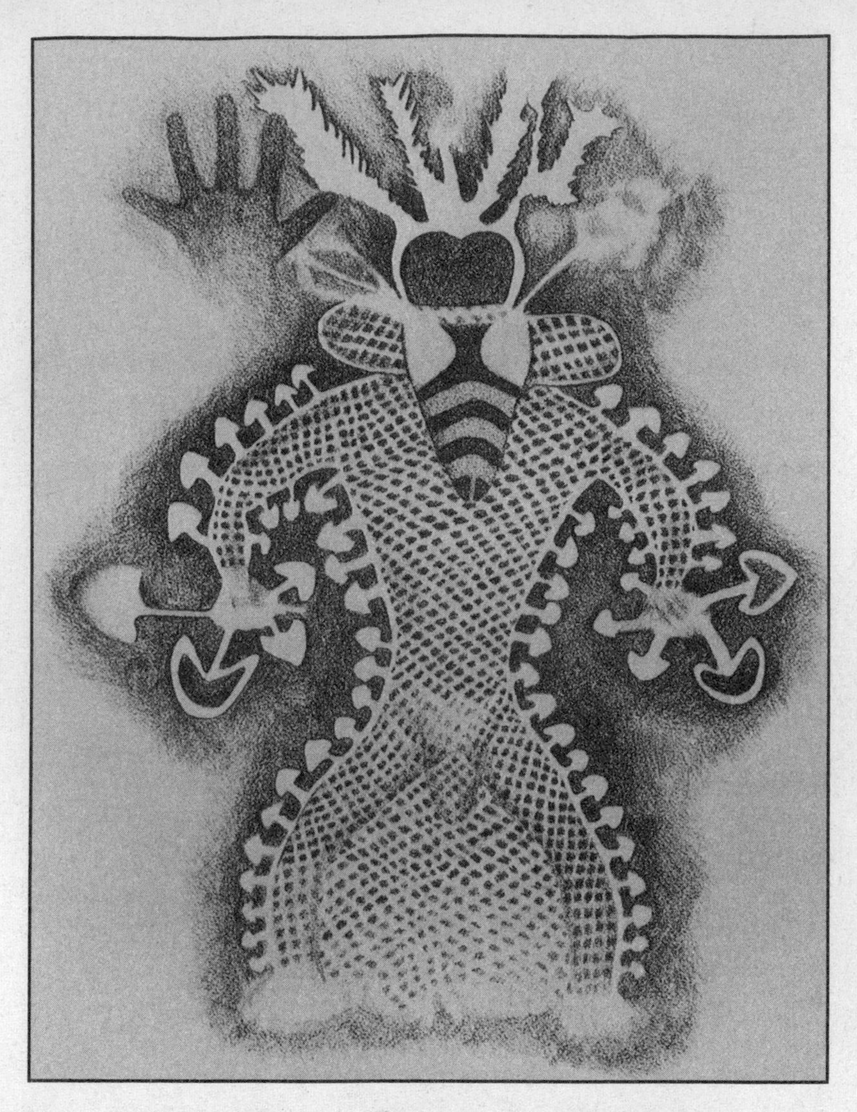

FIGURE 3. The bee-faced mushroom shaman of Tassili-n-Ajjer. Drawing by Kat Harrison-McKenna. From O. T. Oss and O. N. Oeric, *Psilocybin: The Magic Mushroom Grower's Guide*, 1986, p. 71. From the original in Jean-Dominique Lajoux, *The Rock Paintings of the Tassili* (New York: World Publishing, 1963), p. 71.

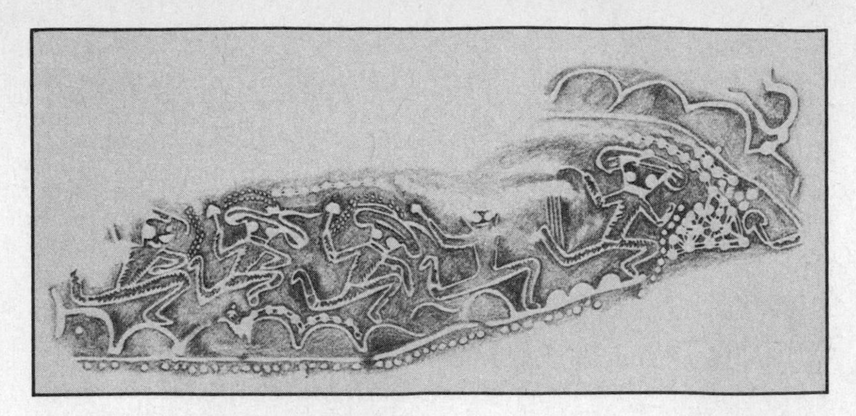

FIGURE 4. Mushroom runners from Tassili. Drawing by Kat Harrison-McKenna. From O. T. Oss and O. N., Oeric, *Psilocybin: The Magic Mushroom Grower's Guide*, 1986, p. 6. From the original in Jean-Dominique Lajoux, *The Rock Paintings of the Tassili*, 1963, pp. 72–73.

sand to seven thousand years ago. Wherever they went, their pastoral lifestyle went with them.[2] The Red Sea was landlocked during much of this time. Lowered sea levels meant that the boot of Arabia was backed up against the African continent. Land bridges at both ends of the Red Sea were utilized by some of these African pastoralists to enter the Fertile Crescent and Asia Minor, where they intermingled with hunter-gatherer populations already present. The pastoral mode had been well established across the ancient Near East by twelve thousand years ago. These pastoral people brought with them a cult of cattle and a cult of the Great Goddess. The evidence that they had such cults comes from rock paintings in the Tassili-n-Ajjer that are from what scholars have named the Round Head Period. This period is named for the style of depiction of the human figure in these paintings—a style not known from any other site.

THE ROUND HEAD CIVILIZATION

The Round Head Period is believed to have begun very early and probably ended before the seventh millennium B.P. Henri Lhote estimates that the Round Head Period lasted several thousand years, placing its beginning somewhere near the start of the ninth millennium. That the Great Goddess was part of the world view of the Round Head–style painters is beyond dispute. A painting from Inaouanrhat in the Tassili includes a wonderful image of a dancing woman (Figure 5). With her outstretched arms and horns extended horizontally on either side of her head, she is the embodiment of the Great Horned Goddess. Her discoverers saw her as having a relationship with the Egyptian Great Goddess Isis, mythical protector of the cultivation of grain.

This impressive figure highlights one of the many problems raised by the Tassili finds. Why, if done at a time when the stratigraphy of the Nile valley shows it to have been nearly deserted, do many of the paintings of the Round Head Period show an unmistakable Egyptian influence in content and style? The logical conclusion is that these motifs and stylistic conceits that we associate with ancient Egypt were first introduced into Egypt by the dwellers of the Western Desert. If proven, this suggestion would indicate the central Sahara as the source of what later became the high civilization of pre-Dynastic Egypt.

PARADISE FOUND?

The Tassili-n-Ajjer of 12,000 B.C. may well have been the partnership paradise whose loss has created one of the most persistent and poignant of our mythological motifs—the nostalgia for paradise, the idea of a lost golden age of plenty, partnership, and social balance. The contention here is that the rise of language, partnership society, and complex religious ideas may have occurred not far from the area where humans emerged—the game-filled, mushroom-dotted grasslands and savannahs of tropical and subtropical Africa. There the partnership society arose and flourished; there hunter-gatherer culture slowly gave way to domestication of animals and plants. In this milieu the psilocybin-containing mushrooms were encountered, consumed, and deified. Language, poetry, ritual, and thought emerged from the darkness of the hominid mind. Eden was not a myth—for the prehistoric peoples of the high plateau of the Tassili-n-Ajjer, Eden was home.

The end of that story may well be the beginning of our own. Is it mere coincidence that at the beginning of the source code of

FIGURE 5. Late Round Head Period painting at Inaouanrhat in the Tassili includes a wonderful image of a dancing Horned Goddess. From *The Search for the Tassili Frescoes* by Henri Lhote, 1959, plate 35, opposite p. 88.

Western civilization, in the Book of Genesis, we read an account of history's first drug bust:

> 3.6. When the woman saw that the fruit of the tree was good to eat, and that it was pleasing to the eye and pleasing to contemplate, she took some and ate it. She also gave her husband some and he ate it. Then the eyes of both of them were opened and they discovered that they were naked; so they stitched fig leaves together and made themselves loincloths.
>
> 3.22. The Lord God made tunics of skins for Adam and his wife and clothed them. He said "The man has become like one of us, knowing good and evil; what if he now reaches out his hand and takes fruit from the tree of life also, eats it and lives forever?" So the Lord God drove him out of the Garden of Eden to till the ground from which he had been taken. He cast him out, and to the east of the Garden of Eden he stationed the cherubim and a sword whirling and flashing to guard the way to the tree of life.

The story of Genesis is the story of a woman who is mistress of the magical plants (Figure 6). She eats and shares the fruits of the Tree of Life or the Tree of Knowledge, fruits which are "pleasing to the eye and pleasing to contemplate." Note that "the eyes of both of them were opened and they discovered that they were naked." At the metaphorical level, they had attained consciousness of themselves as individuals and of each other as "Other." So the fruit of the Tree of Knowledge gave accurate insights, or perhaps it enhanced their appreciation of sensuality. Whichever the case, this ancient story of our ancestors being cast out of a garden by a spiteful and insecure Jehovah, a storm god, is the story of a Goddess-oriented, partnership society thrown into disequilibrium by successive episodes of drought that affected the carrying capacity and climate of the pastoralists' Saharan Eden. The angel with flashing sword who guards the return to Eden seems an obvious symbol of the unforgiving harshness of the desert sun and the severe drought conditions that accompany it.

Tension between male and female is close to the surface in this story and indicates that at the time the story was first recorded the

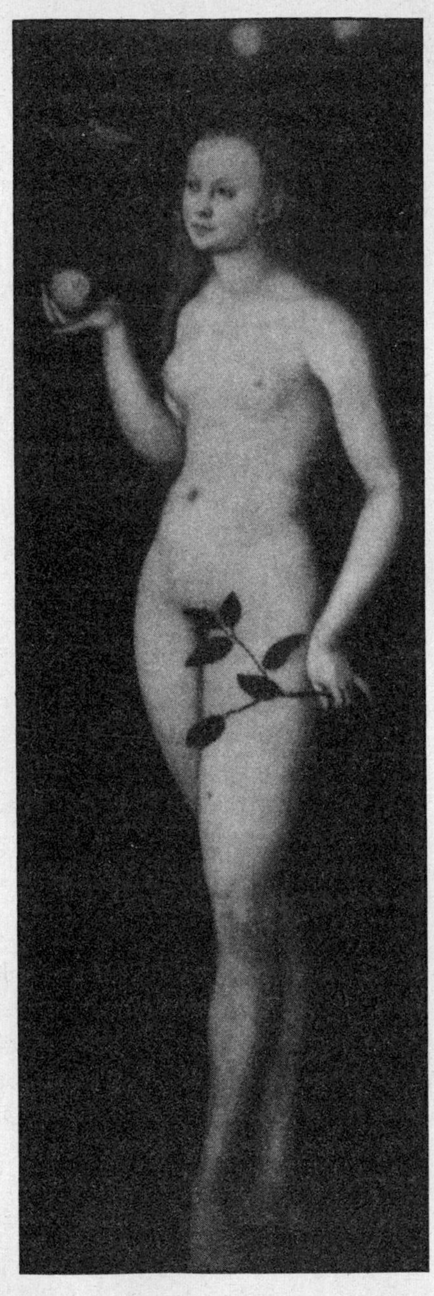

FIGURE 6. *Eve* by Lucas Cranach, c. 1520. Galleria delgi Uffizi of Florence. Courtesy of Fitz Hugh Ludlow Library.

change from partnership- to dominator-style cultures was already well advanced. The woman ate of the fruit of the Tree of Knowledge; this mysterious fruit is the psilocybin-containing mushroom *Stropharia cubensis* that catalyzed the Tassili partnership Eden and then maintained it through a religion that placed a premium on frequent dissolution of personal boundaries into the oceanic presence of the Great Goddess, who is also called Gaia, Geo, Ge, the Earth.

John Pfeiffer, in discussing the Upper Paleolithic cave art of Europe, makes a number of observations that are important for these ideas. He believes that the placement of art within the caves in often nearly inaccessible spots is related to the use of the sites for initiation ceremonies that involved quite complex theatrical effects. He further suggests that what he calls "twilight-state thinking" is a precondition to having great culturally sanctioned truths revealed. Twilight-state thinking is characterized by a loss of objectivity, temporal distortion, and a tendency to experience mild hallucinations, and is nothing more than a gloss for egoless and unbounded psychedelic arousal:

> The prevalence of twilight-state thinking, our very susceptibility to the condition, argues for its evolutionary importance. In extreme cases it results in pathology, derangements and delusions, persisting hallucinations and fanaticism. But it is also the driving force behind efforts to see things whole, to achieve a variety of syntheses from unified field theories in physics to blueprints for utopias in which people will live together in peace. There must have been an enormous selective premium on the twilight state during prehistoric times. If the pressures of the Upper Paleolithic demanded fervid belief and the following of leaders for survival's sake, then individuals endowed with such qualities, with a capacity to fall readily into trances, would out-reproduce more resistant individuals.[3]

Pfeiffer neglects to discuss psychoactive plants and any role they might have played in bringing about twilight thought, and he limits his discussion to Europe. However, the placement of the Tassili rock paintings is similar to that of paintings in many of the European sites, and so it can be presumed that the paintings were used for generally similar purposes; most likely similar religious rites were practiced across southern Europe and North Africa.

The retreat of the glaciers from the Eurasian landmass and the simultaneous acceleration of aridity in the African grasslands eventually brought the "casting out of Eden" allegorically conveyed in Genesis. The mushroom peoples of Tassili-n-Ajjer began to move "east of Eden." And in fact it is possible to trace this migration in the archaeological record.

A MISSING LINK CULTURE

In the middle of the tenth millennium B.C., Palestine, which had been only lightly populated, was the site of the sudden appearance of a remarkably advanced culture that brought with it an explosion in the size of settlements, and in arts, crafts, and technologies, such as had never before been seen in the Near East or, for that matter, anywhere on this planet. This is the Natufian culture, whose crescent-moon flints and elegantly naturalistic carved bonework are unrivaled by anything contemporary found in Europe. As James Mellaart writes, "There is in the Early Natufian a love of art, sometimes naturalistic, sometimes more schematized. The crouching limestone figurine from the cave of Umm ez Zuweitina, or the handle of a sickle from El Wad showing a fawn, are superb examples of naturalistic art, worthy of Upper Paleolithic France."[4] (See Figure 7.)

In spite of the assumption of European academic archaeology that such a culture must have had links with the settlements of Old Europe, the skeletal evidence from Jericho, where the Natufian culture reached its peak, shows clearly that the inhabitants were of Eurafrican stock, fairly robust with long skulls. The ceramic evidence also favors the notion of an African origin: occurring in the Natufian sites is dark, burnished monochrome pottery that is known as Sahara-Sudanese ware. Pottery of this type has been found near the Egyptian-Sudanese border in a situation that suggests that domesticated cattle were present. And it has been found in and near the Tassili-n-Ajjer, evidently having appeared at the end of the Round Head Period. Mary Settegast wrote, "The origin of these African ceramics is unknown. Very recent excavations of Ti-n-Torha

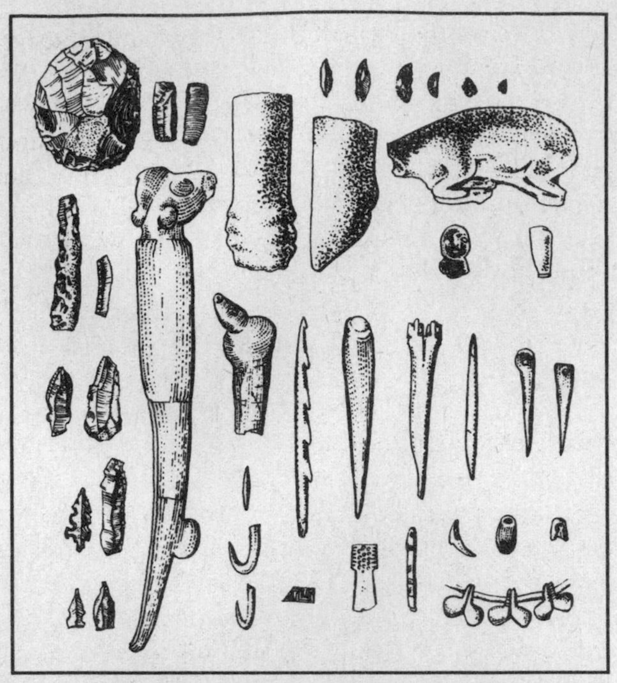

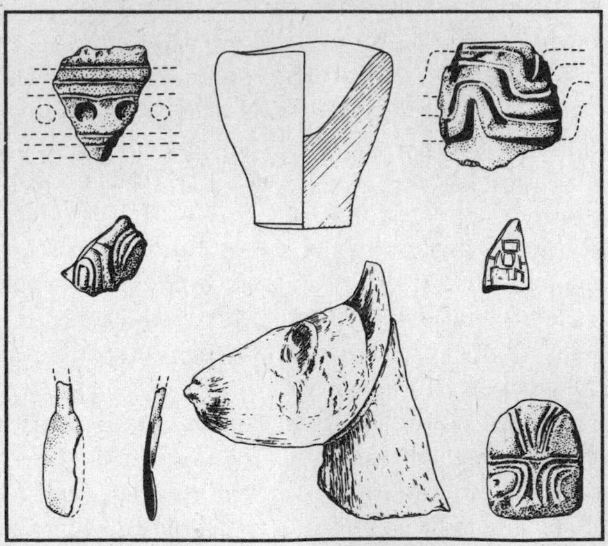

FIGURE 7. Natufian naturalistic art from Palestine. From Figures 5 (p. 25) and 10 (p. 29) of *Earliest Civilizations of the Near East* by James Mellaart (London: Thames & Hudson, 1965).

in the Libyan Sahara have uncovered Sahara-Sudanese type pottery with one carbon-14 reading of 7100 B.C., which, if a reliable date, would suggest a western seniority."[5]

Such statements support the notion that a high culture to the west of the Nile was the source of the new advanced culture appearing in the Nile valley and Palestine.

Of interest in this context is the Natufian culture's particularly close and intense involvement with plants:

> Inquiry into the relationship of environmental and behavioral systems from 10,000–8,000 B.C. reveals that the subsistence base of Natufian populations did not differ appreciably from the local Upper Paleolithic tradition. However, the emphases on plant resources, in the Natufian, allowed for a storable surplus which, in turn, had an effect on the Natufian behavioral patterns. Much of the Natufian material culture (architecture, grinding stones) and settlement pattern were influenced by an intensive exploitation of plant resources.[6]

AFRICAN GENESIS

If the source of the oldest ceramics at the Natufian sites is North African, this would suggest strongly that the Natufian source culture was the previously disrupted partnership paradise that flourished in the wetter and more western regions of the Sahara, especially the Tassili-n-Ajjer. Archaeology may eventually provide answers, but to date no archaeology of significance has been done with these questions in mind. The western Sahara has not been taken seriously as a possible source of the advanced culture that entered Palestine in the mid–tenth millennium B.C. The result of this failure is reflected in comments like the following:

"What is puzzling, however, is that the Palestinian sequence provides nothing at all convincing for an ancestor for the very original first stages of the Natufian. The industry which immediately precedes it . . . is a rather uninteresting culture, having very little in common with its successor. The Natufian, in fact, makes its first appearance apparently full-grown with no traceable roots in the past."[7]

The early Natufians in Palestine settled caves and the terraces in front of caves, and it was in precisely such situations that the rock paintings in the Tassili were done. Further excavation of the major Round Head mural finds in the Tassili might reveal traces of the precocious civilization that is the Natufian source culture.

ÇATAL HÜYÜK

If the Tassili-n-Ajjer can claim consideration as the original Eden and westernmost location of partnership culture, then certainly Çatal Hüyük, in central Anatolia, must be seen as its Neolithic and eastern culmination.

Çatal Hüyük has been called "a premature flash of brilliance and complexity" and "an immensely rich and luxurious city." The stratigraphy for the site begins in the middle of the ninth millennium B.C. Elaboration of cultural forms reaches a pinnacle in Çatal Level VI, in the middle of the seventh millennium B.C. Çatal Hüyük was a huge settlement, spreading over thirty-two acres of the Konya Plain and, at its peak, accommodating over seven thousand people.

Although barely begun, the excavation of Çatal Hüyük has already yielded amazing shrines with cattle bas-reliefs and heads of now extinct aurochs (*Bos primigenius*) covered with ocher designs—the very complex paintings of a complicated civilization (Figure 8). Çatal Hüyük's complexity has puzzled archaeologists:

"Less than three percent of the site has been explored. But Çatal Hüyük has already yielded a wealth of religious art and symbolism that appears to be three or four thousand years ahead of its time. The mature complexity of the traditions at this Neolithic site further presupposes, according to the excavator, an Upper Paleolithic ancestor of whom we have no trace."[8]

I contend that the "Upper Paleolithic ancestor of which we have no trace" is the culture of the Tassili-n-Ajjer. The Natufian culture was a transitional culture directly linking the Round Head culture in Africa with Çatal Hüyük.

In support of this rather startling statement consider the following observations by other scholars. Mellaart said of agriculture at Çatal:

FIGURE 8. Religious shrine at Çatal Hüyük: From *Çatal Hüyük: A Neolithic Town in Anatolia* by James Mellaart (San Francisco: McGraw-Hill Book Co., 1967), Figure 41, p. 128.

> Everything indicates that the plant husbandry of Çatal Hüyük must have a long prehistory somewhere else, in a region where the wild ancestors of these plants were at home presumably in hilly country, well away from the man-made environment of the Konya Plain. . . . The beginnings must be sought in the Natufian of Palestine, the still unknown earlier aceramic of the Anatolian Plateau [of Turkey] and in Khuzistan [farther to the East].[9]

Here is Mellaart on the material culture at Çatal (Figure 9):

> In contrast to other contemporary Neolithic cultures, Çatal Hüyük preserved a number of traditions that seem archaic in a fully developed Neolithic society. The art of wall-painting, the reliefs modelled in clay or cut out of the wall-plaster, the naturalistic representations of animals, human figures and deities, the occasional use of finger impressed clay designs like "macaroni," the developed use of geometric ornament including spirals and meanders, incised

FIGURE 9. A wall painting of insects and flowers, naturalistic in style. The red net pattern has been removed showing insects and flowers. From *Çatal Hüyük: A Neolithic Town in Anatolia* by James Mellaart, 1967, Figure 46, p. 163.

> on seals or transferred to a new medium of weaving; the modelling of animals wounded in hunting-rites, the practice of red-ochre burials, the archaic amulets in the form of a bird-like steatopygous goddess, and finally certain types of stone tools and the preference for dentalium shells in jewelry, all preserve remains of an Upper Paleolithic heritage. To a greater or less extent, such archaic elements are also traceable in a number of other post-Paleolithic cultures, such as the Natufian of Palestine, but nowhere are they so pronounced as in the Neolithic of Çatal Hüyük.[10]

Writing of the painted walls of shrines at Çatal Hüyük, Settegast made this observation:

> The range of pigments used by the Çatal artists was unmatched in the Near East (although equaled or surpassed in the Round Head art of the Sahara). . . . A third type of decoration was accomplished by cutting out silhouettes of animals from the deep accumulations of plaster on the walls, a curious use of interior surfaces that Mellaart [the excavator] believes may have been carried over from the techniques of rock art.[11]

The elegant naturalism of the art at Çatal Hüyük is an echo of the beautiful and sensitive renderings of cattle that typify the Tassili art finds (see, for example, Figure 10). Speaking of the inspired animal art of the Upper Paleolithic, Mellaart says:

> We have already seen a faint survival [of the naturalistic style] in the Natufian of Palestine but it was far more marked in the wall paintings and plaster engravings of the Neolithic site of Çatal Hüyük. There this naturalistic art survived until the middle of the fifty-eighth century B.C., but it is no longer found in the later culture of Hacilar or Can Hasan, cultures which followed in the same area.[12]

What could account for the vitiation of the naturalistic spirit in archaic art that accompanies the changeover from hunting-gathering to agriculture? While the absence of the inspiring mushroom and the visual acuity it imparted cannot be the sole cause, its loss may well have sapped the vitality of the archaic vision. The Goddess-worshiping pastoralists saw more deeply into nature, and their naturalistic style sacrificed esoteric symbolic representation to visual realism, often of the most pristine sort.

The most common motifs at Çatal Hüyük are cattle and bulls and, secondarily, vultures and leopards—all animals of the African grasslands (Figure 11). Of the vultures, Settegast says:

> In any event, if the vulture theme did enter Çatal Hüyük at Level VIII with the pre-dynastic style flint daggers and possibly Sahara-Sudanese-related ceramics, as the excavations to this point suggest, the chance that some of this Anatolian vulture symbolism was actually African cannot be ruled out.[13]

The conclusion that peoples and cultural institutions long old in Africa were entering and flourishing for a time in the Near Eastern environment is logical and difficult to avoid. Mellaart is puzzled that Çatal Hüyük left no great impact on subsequent cultures in the area, remarking that "the neolithic cultures of Anatolia introduced the first beginnings of agriculture and stock breeding and a

FIGURE 10. A beautifully naturalistic rendering of cattle typical of Tassili art, this example from Jabbaren. From Jean-Dominique Lajoux, *The Rock Paintings of the Tassili*, 1963, p. 106.

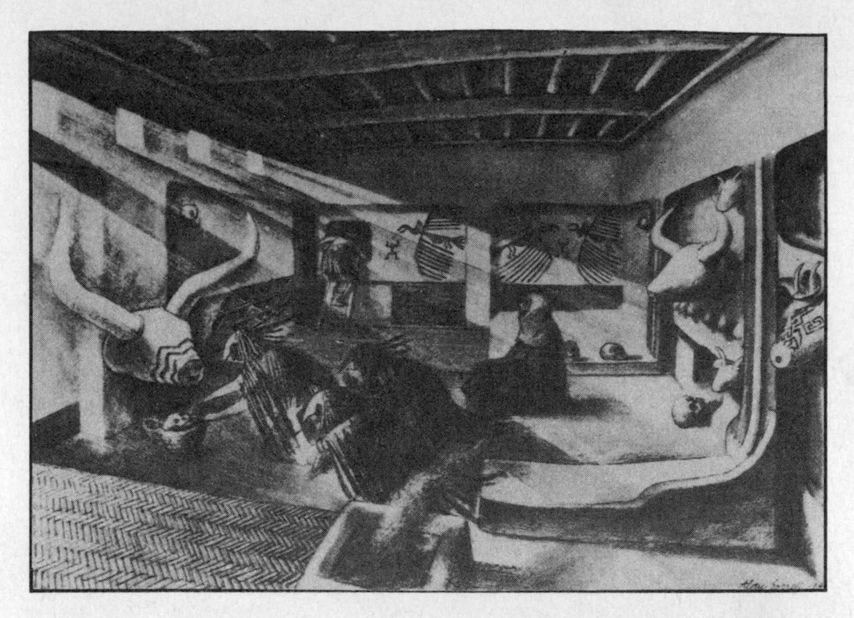

FIGURE 11. The reconstruction of a vulture ritual with priestesses disguised as vultures. From Level VII of Çatal Hüyük, circa 6150 B.C. It is based on the discovery of wall paintings of vultures and skulls found in baskets below each large bull's head on the west and east walls. From *Earliest Civilizations of the Near East* by James Mellaart, 1965, Figure 86, p. 101.

cult of the Mother Goddess, the basis of our civilization."[14] A basis many today still deny, it may fairly be added.

Riane Eisler, who has examined the psychology and mechanisms for maintaining cultural equilibrium in partnership society, argues convincingly that the later pattern to emerge, that of the dominator society, came with the Indo-Europeans—the horse-mounted, wheeled-vehicle cultures from the cold country north of the Black Sea. These are the people of the controversial and hypothetical "Kurgan Waves" of Indo-European population movement. On this matter Eisler's position echoes that of Marija Gimbutas, who wrote:

> The term Old Europe is applied to a pre-Indo-European culture of Europe, a culture matrifocal and probably matrilinear, agricultural and sedentary, egalitarian and peaceful. It contrasted sharply with the ensuing proto-Indo-European culture which was patriarchal, stratified, pastoral,

> mobile, and war-oriented, superimposed on all Europe, except the southern and western fringes, in the course of three waves of infiltration from the Russian steppe, between 4500 and 2500 B.C. During and after this period the female deities, or more accurately the Goddess Creatrix in her many aspects, were largely replaced by the predominantly male divinities of the Indo-Europeans. What developed after c. 2500 B.C. was a mélange of the two mythic systems, Old-European and Indo-European.[15]

Gimbutas believes, in short, that the sedentary matrilinear civilization of Old Europe was disrupted by successive waves of Indo-European invaders with a different culture and language.

Cambridge archaeologist Colin Renfrew has offered an alternative interpretation of this Kurgan Waves theory of Indo-European language diffusion. He claims Çatal Hüyük is the point of origin of the Indo-European language group and the area most likely to be implicated in the invention of agriculture.[16] To support his unorthodox views, Renfrew calls upon the linguistic findings of Vladislav M. Illich-Svitych and Aron Dolgopolsky, which also point toward Anatolia as the home of the Indo-European languages. Dolgopolsky's student Sergei Starostin has argued that about seven thousand years ago Indo-Europeans borrowed a massive number of words from the North Caucasian language of Anatolia. The date of this borrowing argues for our conclusion that Çatal Hüyük was not founded by Indo-Europeans, who would have migrated there during a much later period.[17]

The recent genetic findings of Luigi Cavalli-Sforza and Allan C. Wilson at Berkeley also seem to support this conclusion. The Berkeley group analyzed blood groupings of living populations and traced the populations' genetic roots. They concluded that there is a close genetic relationship among the speakers of the Afro-Asiatic and the Indo-European languages. Their work also supports the view that populations with linguistic roots in Africa had been living on the Anatolian plateau long before the appearance of the Indo-Europeans.

The legacy of Çatal Hüyük was suppressed precisely because of the culture's deep association with the Mother Goddess. The orgiastic psychedelic religion that worshiped the Mother Goddess made the Çatal culture anathema to the new dominator style of warfare and hierarchy. This was a cultural style that arrived suddenly and without warning; the domestication of the horse and discovery of the wheel allowed the Indo-European tribal populations to move south of the Zagros Mountains for the first time. Horse-mounted plunder brought the dominator style to Anatolia, and trampled beneath its hooves the last great partnership civilization. Plunder replaced pastoralism, mead cults finally completed the already well advanced process of supplanting mushroom use; human god-kings replaced the religion of the Goddess.

However, at its height the cult at Çatal Hüyük represented the most advanced and coherent expression of religious feeling in the world. We have very little evidence upon which to reconstruct the nature of the cult acts performed, but the sheer number of shrines in relation to the total number of rooms bespeaks a culture obsessed with religious observances. We know that this was a cult of totemic animals—the vulture, the hunting cat, and always preeminent, the bull or the cow. Later religions in the ancient Middle East were bull worshiping in spirit, but we cannot assume this for Çatal Hüyük. The sculpted heads of cattle that protrude into the cattle shrines at Çatal Hüyük are sexually ambiguous and may represent bulls or cows or simply cattle generally. However, the prevalence of female symbolism in the shrines is overwhelming—for example, the breasts of sculpted stucco that are apparently randomly placed—makes it seem likely that the religious officials were women. The presence of built-in "recliners" in some shrines suggests that curing or midwifery in a shamanic style may have been part of the rites.

It is impossible not to see in the cult of the Great Goddess and the cattle cult of the Late Neolithic a recognition of the mushroom as the third and hidden member of a kind of shamanic trinity. The mushroom, seen to be as much a product of cattle as are milk, meat, and manure, was recognized very early as the physical connection to the presence of the Goddess. This is the secret that was lost some six thousand years ago at the eclipse of Çatal Hüyük.

I am in general agreement with Eisler's view expressed in *The Chalice and the Blade* and hope only to extend her argument by asking the following question: What factor maintained the equilibrium of the partnership societies of the Late Neolithic and then faded, setting the stage for the emergence of the evolutionarily maladaptive dominator model?

In my thinking on this matter, I have been guided by a belief that the depth of the relationship of a human group to the gnosis of the Transcendent Other, the Gaian collectivity of organic life, determines the strength of the group's connection to the archetype of the Goddess and hence to the partnership style of social organization. I base this assumption on observation of shamans in the Amazon, and on observation of the impact of plant hallucinogens on my own psychology and that of my peers.

The mainstream of Western thought ceased to be refreshed by the gnosis of the boundary-dissolving plant hallucinogens long before the close of the Minoan Era, circa 850 B.C. In Crete, and in nearby Greece, awareness of the vegetable Logos continued as an esoteric and diminished presence until the Eleusianian mysteries were finally suppressed by enthusiastic Christian barbarians in A.D 268.[18] The consequence of that severed connection is the modern world—a planet dying under moral anesthesia.

Suppression of the feminine and of knowledge of the natural world has been the hallmark of the intervening centuries. The late medieval Church that conducted the great witch burnings wanted all magic and derangement to be attributed to the Devil; for this reason, it suppressed all knowledge of plants such as thorn apple (*Datura*), deadly nightshade, and monkshood and of the role these plants were playing in the nocturnal activities of the practitioners of witchcraft. And this role was an extensive one; flying ointments and magical salves were compounded out of *Datura* roots and seeds, parts of the plant rich in delirium- and delusion-producing tropane alkaloids. When this material was applied to the witch's body, it produced states of extraordinary derangement and delusion. Hans Baldung's treatment of this subject (Figure 12) leaves no doubt about the terror of the Other the medieval mind projected on the image of intoxicated women. But in the accounts of the Inquisition, the

FIGURE 12. *Compounding the Witches' Unguent*, by Hans Baldung (1514). Mansell Collection. Medieval misogyny on parade. Courtesy of Fitz Hugh Ludlow Library.

central role of plants was never stressed. After all, the Church had no interest in a Devil who is such a diminished figure that he must rely on mere herbs to work his wiles. The Devil must be a worthy foe of the Christos, and hence nearly coequal:

> We must assume that the role of mind-altering plants in some witches' flights was not only under-emphasized, but entirely suppressed for a reason. If this had not been done, then a *natural* explanation for such phenomena would have suggested itself, something in fact advanced by the physicians, philosophers, and magicians quoted here, such as Porta, Weier, and Cardanus. The Devil would then have been left with only a very modest significance, or none at all. If he was assigned nothing more than the role of carnival conjurer, who caused mere illusions to flame up in the heads of witches, he could not have fulfilled the function assigned to him, namely the mighty enemy and seducer of Christendom.[19]

THE VEGETABLE MIND

In view of our present cultural impasse, I conclude that the next evolutionary step must involve not only a repudiation of dominator culture but an Archaic Revival and a rebirth of awareness of the Goddess. Implicit in the ending of profane and secular history is the notion of our involvement with the reemergence of the vegetable mind. That same mind that coaxed us into self-reflecting language now offers us the boundless landscapes of the imagination. This is the same vision of human fulfillment through the "Divine Imagination" that was presciently glimpsed by William Blake. Without such a visionary relationship to psychedelic exopheromones that regulate our symbiotic relationship with the plant kingdom, we stand outside of an understanding of planetary purpose. And understanding planetary purpose may be the major contribution that we can make to the evolutionary process. Returning to the balance of the planetary partnership style means trading the point of view of the egoistic dominator for the intuitional, feeling-toned understanding of the maternal matrix.

A rethinking of the role that hallucinogenic plants and fungi have played in the promotion of human emergence from primate organization can help lead to a new appreciation of the unique confluence of factors responsible and necessary for the evolution of human beings. The widely felt intuition of the presence of the Other as a goddess can be traced back to society's immersion in the vegetable mind. This sense of the female companion explains the persistent intrusion of themes of the mother/goddess even into the most patriarchal domains. The persistence of the cult of Mary in Christianity is a case in point, as is the fervor reserved for the cult of Kali, the destroying mother, and the idea of the divine Purusha in Hinduism. The *anima mundi,* the soul of the world, of Hermetic thought is another image of the Goddess of the World. Ultimately, all of these female images are reducible to the archetype of the original vegetable mind. Immersion in the psychedelic experience provided the ritual context in which human consciousness emerged into the light of self-awareness, self-reflection, and self-articulation—into the light of Gaia, the Earth herself.

GAIAN HOLISM

Deconstruction of dominator cultural values means promotion of what might be called a sense of Gaian Holism—that is, a sense of the unity and balance of nature and of our own position within that dynamic, evolving balance. It is a plant-based view. This return to a perspective on self and ego that places them within the larger context of planetary life and evolution is the essence of the Archaic Revival. Marshall McLuhan was correct to see that planetary human culture, the global village, would be tribal in character. The next great step toward a planetary holism is the partial merging of the technologically transformed human world with the Archaic matrix of vegetable intelligence that is the Transcendent Other.

I hesitate to characterize this dawning awareness as religious; yet that is what it surely is. And it will involve a full exploration of the dimensions revealed by plant hallucinogens, especially those structurally related to neurotransmitters already present in the human brain. Careful exploration of the plant hallucinogens will probe the most Archaic and sensitive level of the drama of the emergence of consciousness: the plant-human quasi-symbiotic relationship that characterized archaic society and religion and through which the numinous mystery was originally experienced. And this experience is no less mysterious for us today, in spite of the general assumption that we have replaced the simple awe of our ancestors with philosophical and epistemic tools of the utmost sophistication and analytical power. Now our choice as a planetary culture is a simple one: Go green or die.

II

PARADISE LOST

7

SEARCHING FOR SOMA: THE GOLDEN VEDIC ENIGMA

Our present global crisis is more profound than any previous historical crises; hence our solutions must be more drastic. Plants and a renewal of our Archaic relationship with plants could serve as the organizational model for life in the twenty-first century, just as the computer operates as the dominant model of the late twentieth century.

We need to think back to the last sane moment that we, as a species, ever knew and then act from the premises that were in place at that moment. This means reaching back in time to models that were successful 15,000 to 20,000 years ago. This shift in viewpoint would enable us to see plants as more than food, shelter, clothing, or even sources of education and religion; they would become models of process. They are, after all, exemplars of symbiotic connectedness and efficient resource recycling and management.

If we acknowledge that the Archaic Revival will be a paradigm transformation and that we really can create a caring, refeminized, ecosensitive world by going back to very old models, then we must admit that more than political exhortation will be needed. To be effective, the Archaic Revival must rest on an experience that shakes each and every one of us to our very roots. The experience must be real, generalized, and discussable.

We can begin this restructuring of thought by declaring legitimate what we have denied for so long. Let us declare Nature to be legitimate. The notion of illegal plants is obnoxious and ridiculous in the first place.

CONTACTING THE MIND BEHIND NATURE

The last best hope for dissolving the steep walls of cultural inflexibility that appear to be channeling us toward true ruin is a renewed shamanism. By reestablishing channels of direct communication with the Other, the mind behind nature, through the use of hallucinogenic plants, we will obtain a new set of lenses to see our way in the world. When the medieval world grew moribund in its world view, secularized European society sought salvation in the revivifying of classical Greek and Roman approaches to law, philosophy, esthetics, city planning, and agriculture. Our dilemma, being deeper, will cast us further back into time in a search for answers. We need to examine the visionary intoxicants of our collective past, which include the strange cult of Soma described in the earliest Indo-European spiritual writings.

No history of plants and peoples could claim completeness without a thorough treatment of the mysterious Soma cult of the ancient Indo-Europeans. As mentioned in Chapter 6, the Indo-Europeans were a nomadic people whose original home has been a matter of scholarly debate and who are associated with patriarchy, wheeled chariots, and the domestication of the horse. Also associated with the Indo-Europeans is a religion based on the magnificently intoxicating Soma.

Soma was a juice or sap pressed out of the swollen fibers of a plant that was also called Soma. The texts seem to imply that the juice was purified by being poured through a woolen filter and then in some cases was mixed with milk. Again and again, and in various ways, we find Soma intimately connected with the symbolism and rituals related to cattle and pastoralism. As will be discussed, the identity of Soma is not known. I believe this connection to cattle is central to any attempt to identify Soma.

The earliest scriptural writings of these Indo-European people are the Vedas. Of these the best known is the *Rig Veda*, best described as a collection of nearly 120 hymns to Soma, the plant and the god. Indeed, the Ninth Mandala of the *Rig Veda* is entirely composed of a paean of praise for the magical plant. The beginning of the Ninth Mandala[1] is typical of the praises of Soma that pervade and typify Indo-European literature of the period:

> Thy juices, purified Soma, all-pervading, swift as thought, go of themselves like the offspring of swift mares; the celestial well-winged sweet-flavored juices, great exciters of exhilaration, alight upon the receptacle.
>
> Thy exhilarating all-pervading juices are let loose separately like chariot-horses; the sweet-flavored Soma waves go to Indra the wielder of the thunderbolt as a cow with milk to the calf.
>
> Like a horse urged on to do battle, do thou who art all-knowing rush from heaven to the receptacle whose mother is the cloud. . . .
>
> Purified Soma, thy celestial steed-like streams as quick as thought are pouring along with the milk into the receptacle; the *rishis*, the ordainers of sacrifice, who cleanse thee, O *rishi*-enjoyed Soma, pour thy continuous streams into the midst of the vessel.[2]

Soma was prominent in the pre-Zoroastrian religion of Iran as "Haoma." "Soma" and "haoma" are different forms of the same word, derived from a root meaning to squeeze out liquid, which is *su* in Sanskrit, and *hu* in Avestan.

No praise seems to have been too excessive to be applied to the magical intoxicant. Soma was thought to have been brought by an eagle from the highest heaven, or from the mountains where it had been placed by Varuna, a member of the early Hindu pantheon. Here is another quote from *Rig Veda:*

> It is drunk by the sick man as medicine at sunrise; partaking of it strengthens the limbs, preserves the legs from breaking, wards off all disease and lengthens life. Then need and trouble vanish away, pinching want is driven off and flees when the inspiring one lays hold of the mortal; the

poor man, in the intoxication of the Soma, feels himself rich; the draught impels the singer to lift his voice and inspires him for song; it gives the poet supernatural power, so that he feels himself immortal. On account of this inspiring power of the drink, there arose even in the Indo-Iranian period a personification of the sap as the god Soma, and ascription to him of almost all the deeds of other gods, the strength of the gods even being increased by this draught. Like Agni, Soma causes his radiance to shine cheeringly in the waters; like Vayu, he drives on with his steeds; like the Acvins, he comes in haste with aid when summoned; like Pusan, he excites reverence, watches over the herds, and leads by the shortest roads to success. Like Indra, as the sought-for ally, he overcomes all enemies, near and far, frees from the evil intentions of the envious, from danger and want, brings goodly riches from heaven, from earth and the air. Soma, too, makes the sun rise in the heavens, restores what has been lost, has a thousand ways and means of help, heals all, blind and lame, chases away the black skin [aborigines], and gives everything into the possession of the pious Arya. In his, the world-ruler's, ordinances these lands stand; he, the bearer of heaven and the prop of earth, holds all people in his hand. Bright shining as Mitra, awe-compelling as Aryaman, he exults and gleams like Surya; Varuna's commands are his commands; he, too, measures the earth's spaces, and built the vault of the heavens; like him, he, too, full of wisdom, guards the community, watches over men even in hidden places, knows the most secret things. . . . He will lengthen the life of the devout endlessly, and after death make him immortal in the place of the blessed, in the highest heaven."[3]

SOMA—WHAT IS IT?

A crucial question arises in any discussion of this powerful plant on whose ecstatic visions all later Hindu religiosity is based: What was the botanical identity of Soma, "the pillar of the World"?

In the nineteenth century this question was nearly impossible to frame. The state of comparative philology was too rudimentary, and there was little impulse to adopt an interdisciplinary approach to the problem: Sanskritists did not talk to botanists, and neither talked to pharmacologists. In fact, to the nineteenth century the question was uninteresting, rather like asking "What song did the Sirens sing?" or "Where is Troy?"

Thanks to the discoveries of Heinrich Schliemann, who followed his own inner voices and promptings, it is generally agreed that we now know where Troy actually stood. And in that spirit of respect for the factual veracity of ancient texts, twentieth-century scholarship has attempted to decipher the botanical identity of Soma. These attempts have ranged from the casual to the exhaustive. The game is precisely the sort that scholars love to play; the answer must be contained in fragmented descriptions in a long-dead language filled with color words and words which only occur once in a literature of a given language. What plant best fits the scattered references to the physical form of this most mysterious member of the visionary flora?

To answer this question we must try to reconstruct the context in which the Indo-Europeans found themselves. One possibility is that the migrations beginning sometime during the sixth millennium B.C. carried the Indo-European tribes far beyond the forest environment suitable to the source of Archaic Soma. Of course, events would have unfolded slowly; the Archaic Soma must have been an item of trade between the original homelands of the Aryans and the frontiers of their southeastward-expanding sphere of influence. Another possibility is that Soma was something that the Indo-Europeans did not come in contact with until they encountered the valley-dwelling pastoralists who presumably used mushrooms and who lived on the Konya Plain of Anatolia. (See Figure 13.)

In either case and over time—as linguistic differences arose, as trade routes became ever longer, and as local substitutes for Soma were experimented with and the local traditions of conquered people assimilated—the original identity of Soma became mingled with myth. Progressively more esoteric, it became a secret teaching, delivered orally and known but to a few, until finally it was forgotten. The preparation of visionary Soma seems to be something that faded as the Indo-European migrations ceased, at a time when reform

FIGURE 13. The double mushroom idol found on the Konya Plain from the Museo di Kayseri. From *Anatolia: Immagini di civilta*, Arnoldo Mondadori, editor, Rome, 1987. Catalog number 99.

and revitalization movements were being strongly felt in Persia as well as on the subcontinent of India.

HAOMA AND ZOROASTER

Perhaps the disappearance of Soma occurred because the newly reform-minded religion of Zoroaster (established circa 575 B.C.), then holding sway on the Iranian plateau, had chosen to take a repressive approach to the ancient sacrament of God-like empowerment. Zoroaster told of Ahura Mazda, a supreme creator, who creates through his own holy spirit and who rules over a world divided between Truth and Lies. The creatures of Ahura Mazda are free and thus responsible for their destiny; the outward symbol of Truth is fire; and the fire altar is the center of Zoroastrian cultic practice.[4] But, as the following makes clear, the old allure of Soma was difficult to suppress:

> There are only two references to Haoma [Soma] in the Gathas [or sacred verses] of Zoroaster, one mentioning Duroaosa "averter of death," and the other alluding to "the filthiness of this intoxicant." These allusions are sufficient to prove that the intoxicating Haoma was under the ban of the great reformer. But in the later Avesta [sacred book of Zoroastrianism], Haoma, like so many other of the old daevas [gods], came back again, and according to Yasna IX–X was in almost every respect the same as the Vedic Soma.[5]

Indeed, Zoroaster may not have actually intended to ban Haoma. Perhaps Zoroaster was merely objecting to the sacrifice of bulls, which was part of the rite. Bull sacrifice would certainly be anathema to anyone who was aware of the connection between cattle and mushrooms in the old religion of the Great Goddess. R. C. Zahner argues persuasively that Zoroaster never abolished the Haoma rite:

> In the *Yasna* the Haoma is prepared for the satisfaction of the "righteous Fravashi of Zoroaster." It is, of course,

> quite true that the Zoroastrians of what we have called the "catholic" period brought back a vast amount of "pagan" material from the older national religion. . . . So far as we can tell, the Haoma rite has been the central liturgical act of Zoroastrianism ever since that religion developed liturgical worship; and the central position it enjoys has never at any time been disputed. This is, however, not true of animal sacrifice; in later times this was practised by some but opposed by others.[6]

What clues are there that might guide us in the search for a botanical identity for Soma? In both the Veda and Avesta, the Soma plant is described as having hanging branches and a yellow color. Its mountain origin is also generally agreed upon. Soma substitutes had to be found once the tradition was forced underground on the Iranian plateau. Presumably, substitutes chosen would be similar in appearance to the original Soma plant. It is also probable that the technical terms of the ritual would be retained, even if the substitute plant did not correspond perfectly to Soma. Since the Soma rite was the essence of Vedic ritual, three daily pressings were necessary to worship the gods, which means large amounts of the plant would have been required. But most important, no plant could substitute for Soma if it were not itself an ecstatic visionary intoxicant worthy of being described in such extravagant terms as these:

> Where there is eternal light, in the world where the sun is placed, in that immortal imperishable world place me, O Soma. . . .
>
> Where life is free, in the third heaven of heavens, where the worlds are radiant, there make me immortal. . . .
>
> Where there is happiness and delight, where joy and pleasure reside, where the desires of our desire are attained, there make me immortal.[7]

HAOMA AND HARMALINE

Attempts to identify Soma have led to heated debates about, for example, the precise meaning of certain color words in the Vedic descriptions.[8] Soma has been identified variously as an *Ephedra*, a plant related to the plant that is the source of the stimulant ephedrine; a *Sarcostemma*, a relative of the American milkweeds; *Cannabis;* and a leafless climber of the genus *Periploca* (see Figure 14). It has also been identified as fermented mare's milk, fermented honey, or a mixture of these and other substances. Recently *Peganum harmala*, the giant Syrian rue, which contains psychoactive substances, has been argued for very persuasively by David Flattery and Martin Schwartz in their intriguing book *Haoma and Harmaline.*[9] They contend that the original identification of Vedic Soma with Syrian rue by Sir William Jones in 1794 was correct. They make their argument using the *Zend Avesta* and other scriptural materials of the Parsi religion that other scholars had passed over. In discussing the ordinarily invisible spiritual world of the after-death state, called *menog* existence in the Avestan religion, Flattery says this:

> The consumption of sauma [Soma] may have been the only means recognized in Iranian religion of seeing into *menog* existence before death; at all events, it is the *only* means acknowledged in Zoroastrian literature . . . . and, as we have seen, is the means used by Ohrmazd when he wishes to make the *menog* existence visible to living persons. In ancient Iranian religion there is little evidence of concern with meditative practice which might foster development of alternative, non-pharmacological means to such vision. In Iran, vision into the spirit world was not thought to come about simply by divine grace or as a reward for saintliness. From the apparent role of sauma in initiation rites, experience of the effects of sauma, which is to say vision of *menog* existence, must have at one time been required of all priests (or the shamans antecedent to them).[10]

THE WASSONS' AMANITA THEORY

Gordon and Valentina Wasson, the founders of the science of ethnomycology, the study of human uses of and lore concerning mushrooms and other fungi, first suggested that Soma might be a

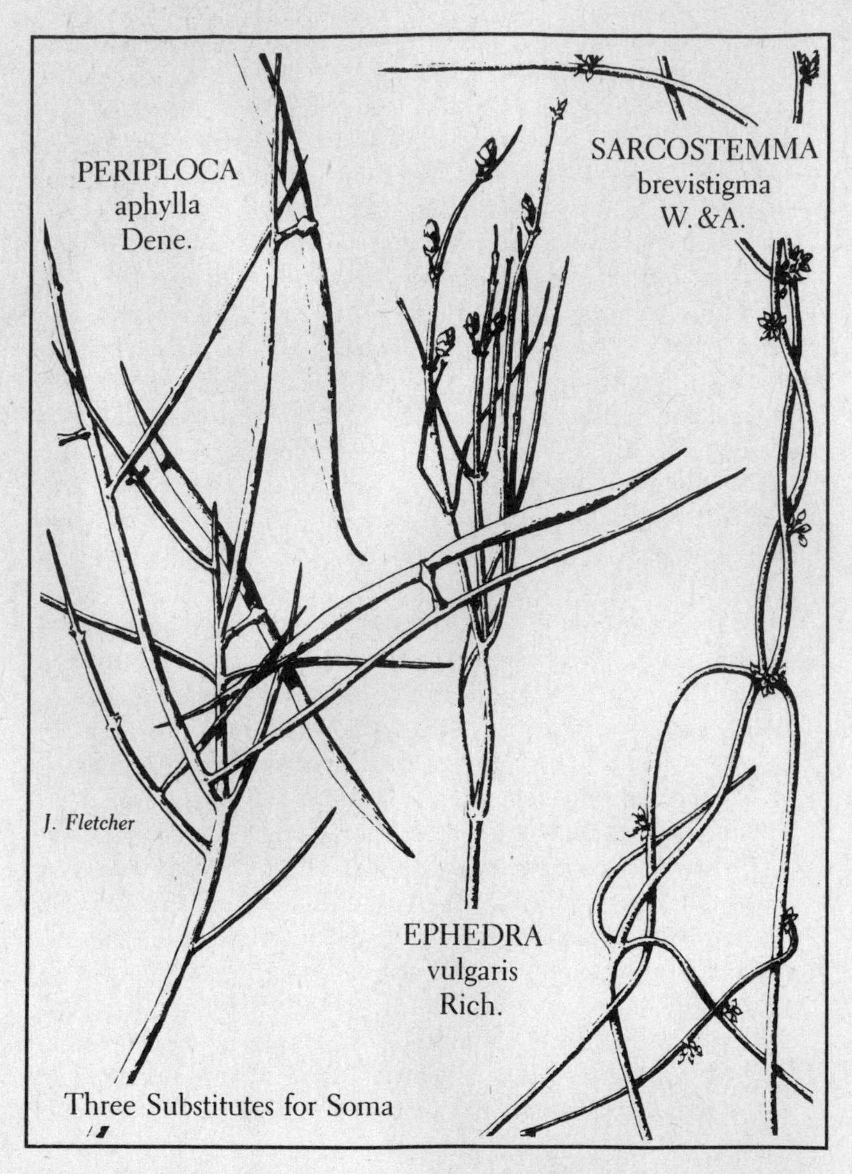

FIGURE 14. Substitutes for Soma. From R. G. Wasson, *Soma: Divine Mushroom of Immortality* (New York: Harcourt Brace Jovanovich, 1971), p. 105.

mushroom—specifically, that it was the scarlet-capped, white-spotted fly agaric, *Amanita muscaria*, an extremely ancient shamanic intoxicant until recently used by the Tungusic tribes of arctic Siberia.

The evidence that the Wassons gathered was massive. By studying the evolution of the languages involved, tracing artistic motifs, and judiciously reexamining and reinterpreting the Vedic material, they made a strong case that a mushroom lay behind the mystery of Soma. Theirs was the first botanically sophisticated, pharmacologically informed inquiry into the identity of Soma.

In other research, the Wassons discovered the existence of still-active shamanic mushroom cults in the mountains of the Sierra Mazateca of Oaxacan Mexico. Gordon Wasson brought samples of Mexican mushrooms to Swiss pharmaceutical chemist and LSD discoverer Albert Hofmann and thus set the stage for the characterization and isolation of psilocybin in 1957. The same psilocybin that I argue was involved in the emergence of human self-reflection on the African grasslands some tens of millennia ago.

In 1971 Gordon Wasson published *Soma: Divine Mushroom of Immortality*. There the case for fly agaric is presented in its most complete form. Wasson was brilliant in advancing the notion that a mushroom of some sort was implicated in the Soma mystery. He was less successful in showing that the species behind the mystery was the fly agaric. He, like all those before him who had attempted an identification of Soma, had forgotten that whatever Soma was, it was a visionary intoxicant of tremendous power and an unparalleled hallucinogen. In contrast, he was well aware that European scholarship had settled upon Siberian shamanism as "exemplary" of all Archaic shamanism and that fly agaric had long been used in Siberia to induce shamanic journeys and initiate neophyte shamans into the fullness of their heritage.

As a result of Wasson's own discoveries in Mexico, it was known that mushrooms other than fly agaric could contain visionary intoxicants, but psilocybin mushrooms were thought to be a strictly New World phenomenon, since no other intoxicating mushrooms were known. Wasson assumed that if a mushroom were Soma, then that mushroom must be a fly agaric. This overemphasis of *Amanita muscaria* has haunted efforts to understand Soma ever since.

Genetically and chemically *Amanita muscaria* is extremely variable; many kinds of fly agaric do not provide a reliable ecstatic experience. Soil considerations and geographic and seasonal factors also affect its hallucinogenic properties. Use of a plant by a shaman does not necessarily mean it is ecstatic. Many rather unpleasant plants are used by shamans to intoxicate themselves and to open the "crack between the worlds." Among these are the *Daturas*—relatives of Jimsonweed, the arborescent *Brugmansias* whose pendulous blossoms are familiar as landscaping ornamentals; bright red and black *Sophora secundifolia* seeds, *Brunfelsias*, and *Virola*-based snuffs made of powdered tree resin. In spite of their shamanic usage, these plants do not induce an ecstatic experience that could inspire the rapturous praise heaped on Soma. Wasson himself was aware that *Amanita* was unreliable, as he himself never had an ecstatic experience from eating *Amanita*.

Instead of realizing that *Amanita muscaria* was an unlikely candidate for Vedic Soma, Wasson became convinced that some method of preparation must have been involved. But no ingredient or procedure has ever been found that reliably transforms the often uncomfortable subtoxic experience of *Amanita* into visionary journeying to a magical paradise. Wasson himself knew of only one inexplicable and unreplicated exception:

> In 1965 and again in 1966 we tried out the fly-agarics (*Amanita muscaria*) repeatedly on ourselves. The results were disappointing. We ate them raw, on empty stomachs. We drank the juice, on empty stomachs. We mixed the juice with milk and drank the mixture, always on empty stomachs. We felt nauseated and some of us threw up. We felt disposed to sleep, and fell into a deep slumber from which shouts could not rouse us, lying like logs, not snoring, dead to the outside world. When in this state I once had vivid dreams, but nothing like what happened when I took the psilocybe mushrooms in Mexico, where I did not sleep at all. In our experiments at Sugadaira [Japan], there was one occasion that differed from the others, one that could be called successful. Rokuya Imazeki took his mushrooms

> with *mizo shiru*, the delectable soup that the Japanese usually serve for breakfast, and he toasted his mushroom caps on a fork before an open fire. When he rose from the sleep that comes with the mushroom, he was in full elation. For three hours he could not help but speak; he was a compulsive speaker. The purport of his remarks was that this was nothing like the alcoholic state; it was infinitely better, beyond all comparison. We did not know at the time why, on this single occasion, our friend Imazeki was affected this way.[11]

The chemical compounds active in *Amanita muscaria* are muscarine and muscimol. Muscarine is highly toxic and like most cholinergic poisons, its activity is reversed by injection of atropine sulfate. Muscimol, the likely candidate for the psychoactivity of the mushroom, has been described as merely an emetic and a sedative.[12] Human exposure to muscimol is not described in the literature. (Incredibly, the obvious step of giving muscimol to human beings to determine its psychedelic potential, if any, has not been undertaken. This fact again points out the queasy illogic that overtakes the academic mentality in the presence of questions revolving around self-induced changes in consciousness.)

To the above let me add my own personal experience of the fly agaric. I have ingested it on two occasions. Once the specimens were a dried collection made at sea level in northern California. My experience of five dried grams was one of nausea, salivation, and blurred vision. Drifting images were present with eyes closed but of a trivial and unengaging sort. My second exposure was a dinner-plate-sized fresh specimen collected at 10,000 feet in the mountains behind Boulder, Colorado. In this case, salivation and stomach cramps were the only effects.

Finally, here is part of an account of fly agaric intoxication by an extremely sophisticated subject, a professional psychotherapist and neurophysiologist. The dose taken was one cup of finely chopped mushroom. The mushrooms came from the Pecos river drainage of New Mexico:

> I was occasionally twitching, a gleam of perspiration over me. Saliva dribbling rapidly out of my mouth. I did not know how the time passed. Thought I was awake, or dream-

ing dreams that were totally lifelike—dreamed in total awareness. I was only dimly or not at all aware of the music being played. Threw off my blanket—very hot sweaty, very cold-chilled, but no visible chills. It seemed unusually quiet inside. I was very stoned. Unlike anything I had felt before—"psychedelic" is too broad a term, too all encompassing, it was not truly psychedelic. It was as if everything were exactly the same but totally unfamiliar—but it all looked like I knew it to be. Except that this world was about a shade (or a quantum level) off—different in an eerie, profound and unmistakable way. I was ataxic [unable to coordinate voluntary movements] and euphoric—there was very little visual stuff.[13]

In short, *Amanita muscaria* is doubtless an effective shamanic vehicle in the floristically limited Arctic environment in which it has been traditionally utilized as a psychoactive agent. But the rapturous visionary ecstasy that inspired the Vedas and was the central mystery of the Indo-European peoples as they moved across the Iranian plateau could not possibly have been caused by *Amanita muscaria*.

WASSON: HIS CONTRADICTIONS AND OTHER FUNGAL CANDIDATES FOR SOMA

Wasson remained convinced that fly agaric was Soma. In his last book, *Persephone's Quest*, published posthumously, he characterized fly agaric as "the supreme entheogen of all time"—apparently on faith, since he admitted it was disappointing and only reported attaining shamanic ecstasy by using psilocybin, which he never introduces into the Soma puzzle. However, he did introduce an interesting caveat when writing of India:

Other fungal entheogens grow at the lower levels. They come in cattle dung, are easily identified and gathered, and are effective. But they fail to conform to Brahmanic practices; they are known to tribals and *sudras* [untouchables].

> Soma on the other hand exacts self-discipline of the priests, a long initiation and training: it is, for proper exploitation, an affair of a priestly *elite*. But the possible role of *Stropharia cubensis* growing in the dung of cattle in the lives of the lower orders remains to this day wholly unexplored. Is *S. cubensis* responsible for the elevation of the cow to a sacred status? And for the inclusion of the urine and dung of cows in the *pancagavya* (the Vedic sacrifice)? And was that a contributing reason for abandoning Soma? Given the ecological conditions prevailing in the Indus Valley and Kashmir, only a few of the Indo-Europeans could know by personal experience the secret of the Divine Herb. The cult of Soma must have been shaped by the peculiar circumstances prevailing in the area, but ultimately those circumstances must have doomed that cult. Today it lives on in India only as an intense and glowing memory of an ancient rite.[14]

In discussing the prohibition against eating mushrooms if one is a Brahmin, a prohibition established in the late Vedic phase, Wasson says:

> We still do not know—we will probably never know—when the proscription came into force, perhaps over centuries while the Vedic hymns were being composed, or possibly when the hierarchs among the Brahmans learned of the entheogenic virtues of *Stropharia cubensis* as known to the lower orders living in India. . . .[15]

Something unusual is going on in these two passages. A great scholar, himself quite a Brahman, an investment banker by profession and an honorary fellow of Harvard University, seems to be behaving in a most unscholarly manner. We know from his own eloquent descriptions that Wasson experienced the ecstasy of psilocybin on more than one occasion. And we know that he never obtained a satisfying experience from *Amanita muscaria*. Yet in these passages he dismisses, ignores, and passes over ample evidence that the mushroom that lay behind the mystery of Soma was the psilocybin-rich *Stropharia cubensis*. He calls it "easily identified" and "effective" but cannot conceive that it could be the Soma he seeks. He asks himself whether *Stropharia cubensis* could have been "a contributing reason for abandoning Soma." Then he ignores his own question. If Soma is *Stropharia cubensis*, then the tradition could be traced unbroken back to prehistoric Africa. Twice in these two passages he refers to "the lower orders," a break with his usual egalitarianism. My contention is that many considerations, some of them unconscious, shaped Wasson's words as he formulated his last statement on the problem that had consumed most of his life.

Those who knew Wasson knew that he had a tremendous aversion to "hippies" and that he was deeply troubled by the events that unfolded in Oaxaca after he published his findings on the mushroom cults that survived there. The predictable migration of adventurers, spiritual seekers, young people, and sensationalists that followed upon Wasson's revelations of the mushroom cults made him bitter and defensive on the subject of psychedelic culture.

> I have often taken the sacred mushrooms, but never for a "kick" or for "recreation." Knowing as I did from the outset the lofty regard in which they are held by those who believe in them, I would not, could not, so profane them. Following my article in *Life* a mob of thrill-mongers seeking the "magic mushroom" descended on Huautla de Jiménez—hippies, self-styled psychiatrists, oddballs, even tour leaders with their docile flocks, many accompanied by their molls. . . . Countless thousands elsewhere have taken the mushrooms (or the synthetic pills containing their active agent) and the chatter of some of them fills the nether reaches of one segment of our "free press." I deplore this activity of the riffraff of our population but what else could we have done?[16]

Wasson maintained a position of stern disapproval of hedonistic use of his beloved "entheogens"—a clumsy word freighted with theological baggage that he preferred to the common term "psychedelic." Perhaps it was this attitude that caused Wasson to decide that his magnum opus, written in colloboration with the French mycologist Roger Heim, *Les Champignons Hallucinogènes du Mexique*, should not be made available in the 1960s in an English translation. There could have been a host of reasons, of course. The fact remains that Wasson's most important work is his only work not available in English.

PEGANUM HARMALA AS SOMA

In fairness to Wasson it must be said that he assumed that *Stropharia cubensis* was first encountered by the Indo-Europeans when they reached India—and thus that it entered the Soma equation rather late. My own contention is that *Stropharia cubensis*, or a conspecific coprophilic species, was well established in Africa, Anatolia, and perhaps the Iranian Plateau millennia before the coming of the Indo-Europeans. This assumption changes the picture in important ways. It means that the invading Indo-European tribes encountered old mushroom-using cultures already in place on the Anatolian and Iranian plateaus.

The increasing dryness of the region could possibly have prompted a search for mushroom substitutes long before the Indo-European invasions. I confess to being impressed by the new data on harmaline put forth by Flattery and Schwartz,[17] arguing conclusively that, at least by late Vedic times, haoma/soma was understood to be *Peganum harmala*. Harmaline, the beta-carboline present in *Peganum harmala*, is distinct in its pharmacological activity from harmine, its near relative which occurs in the South American *ayahuasca* plant, *Banisteriopsis caapi*. It is known that harmaline is more psychoactive and less toxic than harmine. This may mean that *Peganum harmala*, by itself, when brewed to sufficient strength, can give a reliable and ecstatic hallucinogenic experience. It certainly would be true that *Peganum harmala* in combination with psilocybin in any form would synergize and enhance the effects of psilocybin. Perhaps when mushroom supplies were low, this combination was used. Gradually, *Peganum harmala* might have come to supplant the ever-rarer mushrooms altogether. Here is an area where further research is clearly called for.

Whatever the ultimate ethnopharmacological importance assigned to *Peganum harmala,* it is clear that before the Indo-European invasion, the cultures of Anatolia and Iran were of the Çatal Hüyük type. They were cattle-raising, Great Goddess–worshiping partnership societies practicing an orgiastic and psychedelic religion, the roots of which reach back toward Neolithic Africa and the emergence of self-reflecting consciousness.

SOMA AS MALE MOON GOD

The Ninth Mandala of the *Rig Veda* goes into great detail concerning Soma and states that Soma stands above the gods. Soma is the supreme entity. Soma is the moon; Soma is masculine. Here we have a rare phenomenon: a male lunar deity. It is limited to certain North American Indian peoples and to the Indo-Europeans (the German folk conception of the moon is masculine to this day). In studying folklore, the connection between the feminine and the moon is so deep and obvious that a lunar male deity stands out, making its traditional history in any region easy to trace.

In the mythologies of the Near East, there is a lunar god that must have been imported to India from the west. The Babylonian civilization's northernmost outpost was the city of Harran, a city traditionally associated with the original home of Abraham and the beginning of astrology. The patron deity of Harran was a male moon god: Sin or Nannar. He was thought to have arisen from a god of nomads and a protector of cattle related to the masculine cult of the moon god in early Arabia. His daughter Ishtar in time overshadowed all the other female deities, as did her counterpart Isis in Egypt.[18]

As the father, or source, of the Goddess, it is fitting that Sin wears headgear suggestive of a mushroom (see Figure 15). No other deity in the Babylonian pantheon has this headgear. I found three examples of Sin or Nannar on cylinder seals; in each the headgear was prominent, and in one instance the accompanying text by a nineteenth-century scholar mentioned that this headgear was in fact the identifier for the god.[19]

Why was the Harran patron deity connected with the mushroom

FIGURE 15. Cylinder seal that depicts the Harran moon god Sin or Nannar reproduced in *The Dawn of Civilization: Egypt and Chaldea* by Gaston Maspero, 4th ed. (London: Society for Promoting Christian Knowledge, 1922), p. 655. Originally drawn by Faucher-Gudin, from a heliogravure by Ménant, *La Gliptique Orientale*, vol. i. pl. iv., no. 2.

perceived as male? This is a problem for folklorists and mythologists; yet it is clear that the *Stropharia cubensis* mushroom will take the projection of masculinity or femininity with equal ease. It is obviously connected to the moon: it has a lustrous, silvery appearance in certain forms, and the overnight appearance of mushrooms in a field implies that they are active at night when the moon rules the heavens. On the other hand, one can shift the point of view and suddenly see the mushroom as masculine: It is solar in color, phallic in appearance, and imparts a great energy, being traditionally thought of as the child of lightning. The mushroom is most correctly seen as an androgynous shape-shifting deity, which can take various forms depending on the predisposition of the culture encountering it. One can almost say that it is a mirror of cultural expectations, and so for the Indo-Europeans took on a masculine quality and in Saharan Africa and at Çatal Hüyük a very lunar and feminine quality. In any case, it is a hallucinogen or a god that is not wild, that is associated with the domestication of animals and with human culture.

SOMA AND CATTLE

The domestication of the mushroom can serve as the thread to specifically connect the dung-loving *Stropharia cubensis* mushrooms to Soma. That cattle are a major motif in the Soma cult makes little if any sense if one believes that Soma is *Amanita muscaria*. Wasson noted the association of cattle and Soma but went to some lengths to avoid the logical conclusion that Soma must be a dung-loving species: "So much emphasis is laid on cows in the *Rig Veda* and on the urine of bulls in the religion of the Parsis that the question naturally presents itself whether cows consume the fly-agaric and whether they are affected by it, along with their urine and milk. I cannot answer this."[20]

Some eighteen years later, Carl A. P. Ruck, in his contribution to Wasson's last published work, commented on the above passage in a footnote:

> Metaphors of cattle are also attributes of Soma, which can be described as an "udder" that yields the entheogenic milk and as a "bellowing bull," the latter being apparently a characteristic of the mushroom that Perseus picked at Mycenae. The bull is the most common metaphor for Soma, and this manifestation of the sacred plant may underlie the tradition that Zeus, in establishing European civilization, abducted the Anatolian Europea by appearing to her in the form of a bull that breathed upon her the inspiration of the flower he had grazed upon.[21]

In order to save the hypothesis that *Amanita muscaria* is Soma, these authors have seized on the fact that the urine of reindeer and human beings who have eaten *Amanita muscaria* is itself a psychoactive material. Among the Siberian tribes where this fact was noted, the urine is preferred over the plant itself. But *Amanita muscaria* does not grow in grasslands and cattle do not habitually graze on mushrooms, nor is there any reason to believe that if they did their urine would have psychoactive properties, as the hallucinogens would probably have been metabolized.

WASSON'S DOUBTS

Wasson himself was not as certain as his published statements might seem to indicate. In 1977 Wasson wrote the following in answer to my inquiry concerning the *Stropharia* versus *Amanita* question:

> Your question about *Str*[*opharia*] *cubensis* has also bothered me. When Roger Heim and I went to India in 1967, in the Simlipal Hills of Orissa, I was given an account of a mushroom growing in cow's dung that tallied perfectly with *Str. cubensis* even to its psychoactive powers. My informant said that everyone avoided it. He seemed not to be withholding anything. He said he would deliver the mushroom to us, but though we stayed there a couple more days, I saw no more of him. Our purpose in going to India was altogether different. It will be necessary to pursue *Str. cubensis* further not only in India but elsewhere in the world. Of course *Str. cubensis* must flourish in India. Did it play a part in the abandonment of Soma? Inebriation from *Str. cubensis* and the other psilocybin species is clearly, in my opinion, superior to A[*manita*] *muscaria*. I may develop this as one of several further ideas that I propose to include in my book next after this one, which I am now drawing to a close.[22]

Eventually, however, Wasson contradicted this position.

A MORE PLAUSIBLE ARGUMENT

As the arguments for *Amanita* being Soma are quite tortured, I believe the idea is best abandoned. The web of textual and linguistic associations that was so convincing to some can probably not be saved. Nevertheless, a more plausible reconstruction might run as follows:

In their original homeland north of the Black Sea, the Indo-Europeans might well have practiced a shamanic religion with close similarities to the *Amanita*-using shamanism characteristic of the

Koryak, Chukchi, and the Kamchadal peoples of northeast Siberia. The Indo-Europeans were at that time surrounded to the north and the east by Finno-Ugric peoples with a presumed long history of fly agaric use.

In the sixth millennium B.C., settled agricultural populations had already been present in Europe for over two thousand years and urban civilizations were already ancient in the fertile river valleys of the Near East and the Anatolian plain. Sometime in this millennium, the first extensive Indo-European colonization of the Asian steppes and desert areas began. On the Eurasian steppes north of the Black Sea, the Caucasus, and the Taurus and Zagros mountains, the horse provided the key. If the domestication of cattle in Africa had set the stage for mushroom-using, Goddess-worshiping partnership societies, among the Indo-Europeans the domestication of the horse reinforced mobility, male dominance, and a social economy based on rape and plunder. Wheeled vehicles, first invented on the fringes of the Caucasus where woodland and steppe meet, soon spread among the Indo-European tribes. With horse and chariot, they began moving west into the zone of established farming groups, east into central Asia, and south toward Lake Van, where they encountered the urban cultures of the Anatolian and Iranian plateaus. These were cultures long in place and connected to a past that reached south and west, to the cradle of consciousness in temperate African grasslands. Use of psilocybin was a folk practice as old as these cultures themselves.

THE INDO-EUROPEANS

Whatever relationship the Indo-Europeans may have had to *Amanita* in their region of origin, it is most reasonable to suppose that the Vedas were written during the long centuries of their migrations toward the Indian subcontinent. These were centuries during which the Indo-Europeans subjugated and assimilated the valley pastoralists whom they conquered. From their contact with these cultures, the Indo-Europeans for the first time encountered the miracle of Soma and the awesome power in psilocybin. And while the Great Mother Goddess was suppressed in favor of the early Vedic pantheon, and the partnership pattern was replaced by male dominance

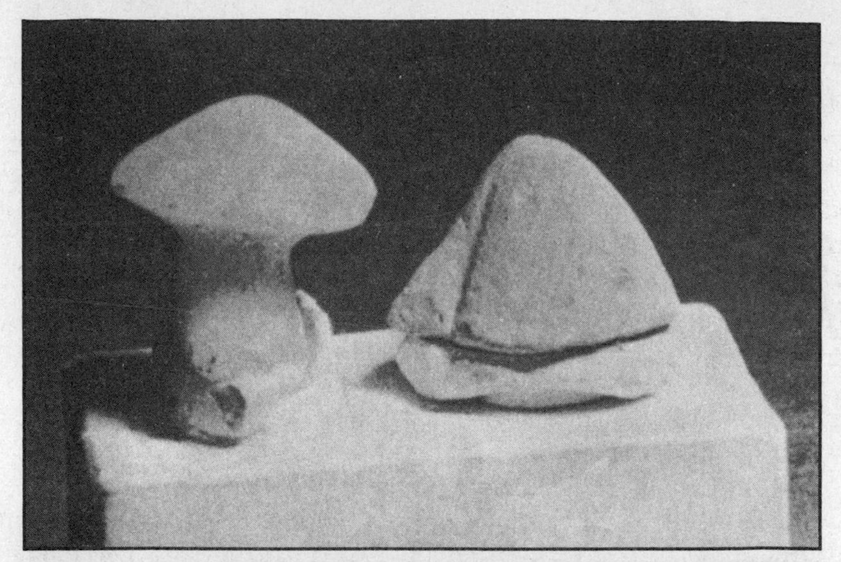

FIGURE 16. Green mushroom-shaped stones of the Vinca site from Marija Gimbutas's *The Goddesses and Gods of Old Europe* (Berkeley: University of California Press, 1982), Figures 223 and 225.

and patriarchy, still what was retained, exalted, and deified during the nomadic phase was the mushroom, now become Soma, Thunderbolt of Indra.

And though in earlier chapters, I argued for psilocybin use in prehistoric Africa and Asia Minor, the evidence for this position is pictorial and circumstantial; it is not yet direct. A remarkable 2,500-year-old vessel with two grinning anthropomorphic mushrooms embossed on its surface found in Anatolia suggests that physical evidence of Middle Eastern mushroom use may soon emerge. (See Figure 13.) Small mushroom-shaped objects carved from green stone have been found in Yugoslavia as well.[23] (See Figure 16.)

As climatic conditions changed and as the Indo-Europeans migrated farther and farther east, it is likely that the mild temperatures and grassland conditions required by *Stropharia cubensis* ceased to be available. Other mushrooms may have enjoyed use as Soma substitutes, and of these *Amanita muscaria* may have been preferred, because of its availability in colder climes, its psychoactivity (however ambiguous), and its striking appearance.

There are a number of possible problems with this theory. Primary among them is the lack of confirmation of the presence in India of *Stropharia cubensis* or other psilocybin-containing mushrooms. *Amanita muscaria* is also rare in India. I predict, however, that a careful search of the flora of India will reveal *Stropharia cubensis* as a common indigenous component of the biome of the subcontinent. The desertification of the entire area from North Africa to the region around Delhi has distorted our conception of what occurred when ancient civilizations were in their infancy and the area received higher rainfall.

The psilocybin mushroom religion, born at the birth of cognition in the grasslands of Africa, may actually be the generic religion of human beings. All later adumbrations of religion in the ancient Near East can be traced to a cult of Goddess and cattle worship, whose Archaic roots reach back to an extremely ancient rite of ingestion of psilocybin mushrooms to induce ecstasy, dissolve the boundaries of the ego, and reunite the worshiper with the personified vegetable matrix of planetary life.

8

TWILIGHT IN EDEN: MINOAN CRETE AND THE ELEUSINIAN MYSTERY

In the absence of a partnership community and with the loss of the psychoactive plants that catalyze and maintain partnership, nostalgia for paradise appears quite naturally in a dominator society. The abandonment of the original catalyst for the emergence of self-reflection and language, the *Stropharia cubensis* psilocybin-containing mushroom, has been a process with four distinct stages. Each stage represents a further dilution of awareness of the power and the numinous meaning resident in the mystery.

The first step away from the symbiosis of the human-fungal partnership that characterized the early pastoralist societies was the introduction of other psychoactive plant substitutes for the original mushroom. This psychoactivity can range from being equal in the depths of its profundity to the *Stropharia cubensis* psilocybin intoxication, as in the case with the classical hallucinogens of the New World tropics, to being relatively trivial. Examples of the latter are the use of *Ephedra*, a stimulant, and fermented honey as Soma substitutes.

In the case of *Stropharia cubensis* in Africa, a gradual trivialization scenario is reasonable: With changes in climate, frequent, if not continual, low levels of mushroom ingestion gradually gave way to use that was merely seasonal. Conscious ceremonial use of mushrooms must have been at its apex during this seasonal availability phase, which may have lasted many thousands of years. Gradually, as mushrooms and mushroom ecologies grew more rare, there may have been efforts to preserve mushrooms by drying and by preserving them in honey. As honey itself easily ferments into an alcoholic intoxicant, it is possible that over time a practice of mixing fewer and fewer mushrooms in more and more honey may have encouraged replacement of the mushroom cult with a cult of mead. No greater shift of social values is possible to imagine than that which would accompany the gradual changeover of a psilocybin cult to an alcohol cult.

Such gradual profanation of a psychoactive plant sacrament merges easily into the second step in the abandonment of the original psychosymbiotic mystery; the second step is the substitution of completely inactive materials for active ones. In this situation, the substitutes, though usually still plants, are really no more than symbols of the former power of the mystery to authentically move initiates.

And in the third stage of the process, symbols are all that is left. Not only are psychoactive plants now out of the picture, but plants of any sort have disappeared, and in their place are esoteric teachings and dogma, rituals, stress on lineages, gestures, and cosmogonic diagrams. Today's major world religions are typical of this stage.

The third stage leads into yet another stage. This other stage is, of course, the complete abandonment of even the pretense of remembering the felt experience of the mystery. This last stage is typified by secular scientism as perfected in the twentieth century.

We could perhaps even posit a further aspect of this fourth stage in the process of abandonment: the rediscovery of the mystery and its interpretation as evil and threatening to social values. The current suppression of psychedelic research and the hysteria fanned by pharmaphobic media is an obvious case in point.

The discussion of Minoan civilization and the mystery cults it spawned and sheltered takes us to the domain of the plant substitutes for *Stropharia cubensis* psilocybin. These were powerful cults with powerful plants to aid in the formulation of a religious ontology—but in all likelihood they were not directly dependent on sources of psilocybin for the attainment of ecstasy. In Minoan Crete and still later at Eleusis on the Greek mainland, hallucinogenic indoles of other types were admitted as techniques of ecstasy. Cultural and climatic conditions made the original source of the boundary-dissolving psilocybin ecstasy no more than a memory and its image no more than a symbol.

THE FALL OF ÇATAL HÜYÜK AND THE AGE OF KINGSHIP

James Mellaart, the principal investigator of the site, makes the point that for all its brilliance Çatal Hüyük had no impact on the societies around it. A disastrous series of fires swept through Levels V and VIA around 6500 B.C., and the city was abandoned, making all too clear that the age of undefended cities, the age of partnership, was ending. From then on, partnership-based social institutions and the old Mother Goddess religion in the Near East would witness a slow erosion and fragmentation. Refugees from the fall of the Çatal Hüyük civilization were scattered. Some few of them fled to the island of Crete:

> The story of Minoan civilization begins around 6000 B.C.E., when a small colony of immigrants, probably from Anatolia, first arrived on the island's shores. These immigrants brought the Goddess with them, as well as an agrarian technology that classifies these first new settlers as Neolithic. For the next four thousand years there was slow and steady technological progress, in pottery making, weaving, metallurgy, engraving, architecture, and other crafts, as well as increasing trade and the gradual evolution of the lively and joyful artistic style so characteristic of Crete.
>
> On the island of Crete where the Goddess was still supreme there are no signs of war. Here the economy prospered and the arts flourished. And even when in the fifteenth century B.C. the island finally came under Achaean do-

minion—when the archaeologists no longer speak of Minoan but rather of a Minoan-Mycenaean culture—the Goddess and the way of thinking and living she symbolized still appear to have held fast.[1]

The ambience of Minoan-Mycenaean religion was one of realism, a sense of the vitality of *bios*, and sensual celebration. The snake-handling Minoan nature Goddess is representative of all these values. In all Minoan depictions, her breasts are full and bare and she handles a golden snake. Scholars have followed shamanic convention and have seen in the snake a symbol of the soul of the deceased. We are dealing with a goddess who, like Persephone, rules over the underworld, a shamaness of great power whose mystery was already millennia old.[2]

Meanwhile, on the mainland of Asia Minor, the Indo-European waves of successive migration abated and the great urban river valley civilizations arose. Kingship and chariot warfare and the travails of great male heros now held sway in the collective imagination. Warfare and the building of fortified cities had become the enterprise of civilization. In the age of kingship, only Crete—an island and in those times remote from the events of Asia Minor—harbored the old partnership model.

The mysterious Minoan civilization became the inheritor of the style and gnosis of forgotten and far-off times. It was a living monument to the partnership ideal, enduring for three millennia after the triumph of the dominator style was everywhere else complete.

MINOAN MUSHROOM FANTASIES

The question naturally arises of the relationship of Minoan society to the Archaic source of power behind the partnership ideal, namely, mushroom psilocybin. Was the old mushroom religion of the African Eden preserved and absorbed into the life of Minoan culture? Did the people still search for ecstasy but through other means in the absence of mushrooms?

What are we to make of the worship of pillars that characterized Minoan religion, remembering that Soma was called "pillar of the World" in the *Rig Veda*? It is generally assumed that these pillars are related to the Great Goddess religion and her vegetation cult, but could they be explicit echoes of the memory of the mushrooms?

> The palaces were characteristic of the style of Minoan culture and probably were sacred in their totality, though only certain rooms were employed in the cult. . . . On the upper floors we find several rooms each with a single round column in the center, a column broadening toward the top, as—to cite a simple example—in the so-called temple tomb near the palace of Knossos. The religious implications of this column cannot be doubted.[3]

Was the pillar somehow an esoteric reference to the mushroom mystery, or a last aniconic vestige of the mushroom image?[4] Such columns were widely understood to stand for a sacred tree. The column was connected with images and rituals of vegetative significance that were very old. Was the use of mushrooms on Crete once an active and widespread cult, or was mushroom use only a memory of long-forgotten times before the arrival of the Goddess worshipers on Cretan shores? The great mystery cults that coexisted in the ancient Greek world of the fourth century B.C., which we call Dionysian and Eleusinian, were the last frail outposts in the west of a tradition of using psychoactive plants to dissolve personal boundaries, and to gain access to gnosis; true knowledge of the nature of things, that was many thousands of years old. Although they can be traced to Cretan origins, it is not clear whether psychoactive substances were a part of the celebration of the Minoan rites for the Goddess. Archaeological evidence on this point is lacking. Strong cultural evidence, however, to be discussed below, suggests that Eleusis, that most Greek of all the Mysteries, was a cult of plant-induced group psychedelic ecstasy.

A curious and suggestive myth may shed some light on the problem of psychoactive plant use in the Minoan-Mycenaean context. This myth, the story of Glaukos, son of King Minos and Pasiphaë, the Moon Goddess, has received little attention from modern scholars. It is preserved in complete form in only two late sources, Apollodorus and Hygeinus; fragmented versions are found in earlier writers.[5] Parts of the story also appear in Aeschylus's *Kressai*, Sophocles' *Manteis*, and Euripides' *Polyidos*. The fascination this myth had for the great dramatists suggests that it was a popular theme of the Classic period. The story is old, definitely from the prehistoric phase of Greek mythological thinking. The retelling below follows the version of Apollodorus.[6]

THE MYTH OF GLAUKOS

While Glaukos, the son of Minos and Pasiphaë, was still a small child, he died from falling into a jar, a *pithos*, filled with honey, while he was pursuing a rat, or a fly; the manuscripts are uncertain. Upon his disappearance his father Minos made many attempts to find him, and finally went to diviners for advice on how he should go about his search. The *Kouretes* answered that Minos had among his herds a cow of three different colors and that the man who could offer the best simile for this phenomenon would also be the one to know how to restore the boy to life. The diviners gathered together for this task, and finally Polyidos, son of Koiranos, compared the cow's colors to the fruit of the bramble. Compelled thereupon to search for the boy, he eventually found him by means of his powers of divination, but Minos next insisted that Polyidos must restore the boy to life. He was therefore shut up in a tomb with the dead body. While in this great perplexity, he saw a snake approach the corpse. Fearing for his own life should any harm befall the boy's body, Polyidos threw a stone at the serpent and killed it. Then a second snake crept forth, and when it saw its mate lying dead it disappeared, only to return with an herb which it placed on the dead snake, immediately restoring it to life. After Polyidos has seen this with great surprise, he took the same herb and applied it to the body of Glaukos, thereby raising him from the dead. Now although Minos had his son restored to life again, he would not allow Polyidos to depart home to Argos until he had taught Glaukos the art of divination. Under this compulsion Polyidos instructed the youth in the art. But when Polyidos was about to sail away, he bade Glaukos spit into his mouth.

This Glaukos did, and thereby unwittingly lost the power of divination.

This must suffice for my account of the descendants of Europa.[7]

Let us attempt to analyze this peculiar story. First of all, it is necessary to comment on the significance of the names of the two main characters: Polyidos is clearly "the man-who-has-many-ideas," and Glaukos simply means "blue-gray." The meaning of Glaukos was for me the entry point into the intention of the myth. It is well known among mycologists that the flesh of *Stropharia cubensis* and other psilocybin mushrooms has the property of staining a bluish color when bruised or broken. This blue staining is an enzymatic reaction and a fairly reliable indicator of the presence of psilocybin. Glaukos, the youth who is preserved in the jar of honey, seems symbolic of the mushroom itself. Indeed, Wasson mentions the frequent allusions to honey in connection with Soma in the *Rig Veda*. He rejects the notion that mead, the fermented form of honey, could have been the basis of Soma: "Honey, *madhu*, is mentioned frequently in the *Rig Veda* but mead never. Honey is cited for its sweetness and also is often applied as a metaphor of enhancement to Soma. There is reason to think it was used on occasions to mix with Soma, but the two were never confused."[8]

HONEY AND OPIUM

The antiseptic properties of honey have made it a preferred medium among many peoples for the preservation of delicate foods. And in Mexico honey has long been used to preserve psilocybin-containing mushrooms. The fact that Glaukos, the blue-gray one, fell into a honey pot (whose shape suggests the bucket-shaped graves of the Natufians) and was preserved there until the time of his resurrection seems highly suggestive. Herodotus mentions that the Babylonians preserved their dead in honey, and the use of large storage vessels, or *pithoi*, for burying the dead was widespread in the Bronze Age Aegean. The motif of cattle is present in the story in the bizarre section concerning the simile of the three-colored cow and the need to demonstrate linguistic facility as a precondition to being able to find the lost child. And the serpent, familiar from the Genesis story of Eden, makes a cameo appearance—and once again proves to have accurate and secret information concerning plants, especially plants that confer immortality. Polyidos, the shaman figure, uses the information gained from the serpent to return Glaukos to life; he shares his shamanic understanding with the boy, but later, all the information leaves Glaukos and returns to his departing teacher. This may refer to the elusive nature of the visions glimpsed during mushroom intoxication.

The story is obviously garbled in this version, and the simile contest regarding the three-colored cow hardly makes sense; yet all the motifs of a barely remembered mushroom cult are there—themes of death and rebirth, cattle, serpents with herbal knowledge, and a blue-gray child who is preserved in honey. A parallel example is provided by the mushroom cults of the New World: throughout their range in Mesoamerica, the psychoactive mushrooms are thought of as small children—*los niños,* "the dear sweet little ones," Maria Sabina, the mushroom shamaness of Huautla de Jiménez, called them. This is an instance of the motif of the alchemical children, the elfin denizens of some nearby magical continuum, accessed through psilocybin.

We may never know with certainty the role that hallucinogenic fungi and plants had in the Minoan world. Much can change over the length of nearly four thousand years, and we know from the scholarship of Kerényi and others that the late Mycenaean-Minoan civilization was more fascinated by opium than by psychedelic plants:

> It may be presumed that toward the end of the late Minoan period, opium stimulated the visionary faculty and aroused visions which had earlier been obtained without opium. For a time, an artificially induced experience of transcendence in nature was able to replace the original experience. In the history of religions, periods of "strong medicine" usually occur when the simpler methods no longer suffice. . . . Opium was consonant with the style of Minoan culture and helped to preserve it. When Minoan culture came to an end, the use of opium died out. This

culture was characterized by an atmosphere which in the end required such "strong medicine." The style of Minoan *bios* is discernible in what I have called the "spirit" of Minoan art. This spirit is perfectly inconceivable without opium.[9]

The openness of Minoan society to the inclusion of opium in its religious rites is indicative of a willingness to associate ecstasy and the pursuit of altered states of consciousness with plant alkaloids. It is therefore a strong argument that other plants may have been utilized originally.

THE DIONYSUS CONNECTION

Dionysus, son of Zeus and the mortal Semele, twice born, god of intoxication who brings madness to women, has never been a comfortable figure in the Greek pantheon. There is something older, wilder, and strange that hovers about him. He is a vegetation god, a mad god and a dying god, a god of orgy, androgyny, and intoxication—and yet more, for from his miraculous birth onward, his story contains unique elements. Dionysus was twice born because his mother died, consumed in a lightning storm before she could give birth:

> The father did not let his son perish. Cooling ivy tendrils protected him from the heat by which his mother was consumed. The father himself assumed the role of the mother. He took up the fruit of the womb, not yet capable of life, and placed it in his divine body. And when the number of months was accomplished, he brought his son into the light.[10]

This notion of the "twice born god" anticipates the mystery of the Christos in ways scholarship has not fully explored. Only in the late phase of Greek culture was Dionysus transformed into the god of wine and drunken revelry; the older stratum of material is darker and touched with the bizarre.

Semele was thought to be one of the four daughters of King

Cadmus of Thebes, according to Graves.[11] A clue to the Minoan connections surrounding Dionysus is the fact that Semele, though mortal, was accorded her own special cult honors as a goddess. The rites of Dionysus as practiced on the island of Myconos were deeply entangled with rites that honored his mother. Scholars have in fact reconsidered Semele's mortality and decided that she may have been a goddess all along. Kretschmer pointed out that Apollodorus equated Semele with Ge, the Thracian form of Gaia.

In the older stratum, the Minoan stratum, Dionysus is the son of the Great Mother Goddess and is totally subservient to her. A point of view sensitive to the polarity of the partnership versus dominator relationship in the ancient world and of the change from one to the other cannot fail to see this as an important clue. Is not Dionysus, in his androgyny, in his madness, in his personification of ecstatic intoxication, the image of the spiritual crises that overcame the Minoan Archaic ideal? A male god, but softened by the androgynous values of Gaian culture, a dying god, personifying the death agony of the symbiotic relationship to vegetation that male dominance, Christianity, and the phonetic alphabet would finally overthrow. A god comprehensible only to initiates, usually women, in the cult—and from the point of view of the patriarchy, something wild, ancient, and potentially dangerous.

The Dionysian theme entered staid Greece from the south, from island cultures with roots ten thousand years deep in the religion of the mushroom Mother Goddess: it entered from Asia Minor, but via four millennia of incubation within Minoan civilization. The mysteries that were planted on Grecian shores at Eleusis were the latest, last, and most baroque adumbrations of the great Archaic religion of the Goddess, cattle, and ecstatic intoxication by indole hallucinogens.

THE MYSTERY AT ELEUSIS

Each September, for two thousand years that more than span the classical Greek and Roman civilizations, a great festival was celebrated on the Eleusinian plain near Athens. In that place, tradition held, the goddess Demeter had been reunited with her daughter, Kore or Persephone, who had been abducted into the nether world by its lord and ruler, Pluto. These two goddesses, seeming sometimes more sisters than mother and daughter, are the two great figures around which the Eleusinian Mysteries were celebrated. The festival of the Mysteries was held on two occasions during the Athenian year: the Lesser Mysteries celebrated in the spring to welcome the return of vegetation anticipated the Great Mysteries celebrated at harvest time. The Mysteries were clearly connected to Minoan rites:

> The oldest Telesteria [cult structures] are pre-Hellenic; the name Eleusis suggests pre-Hellenic Crete; certain cult vessels, the *kernoi*, and libation jugs are common to Eleusinian and Minoan cults; the form of the Telesteria may possibly be a further development of the so-called Minoan theater; the *anaktoron* is the same as the Cretan repositories and so-called house chapels; the purifications of the Eleusinian cult come from Crete, where they originally belonged to the Minoan religion; the kernel of the mysteries is a cult of fertility, which is also the kernel in the Minoan religion; a double ancient tradition traces the mysteries to Crete: on the one hand Diodoros, who stands independently, on the other the Homeric *Hymn to Demeter*. . . . These conclusions, established nearly twenty years ago, have since been adopted by leading historians of religion. The correctness of the interpretation, achieved without the more intimate knowledge of the basic content of the Minoan religion which we now have, is further strengthened by the present research.[12]

Though Eleusis has commanded the attention of many scholars, we still do not have a definitive understanding of exactly what it was that gave the Mystery a power over the Hellenistic imagination such that for nearly two thousand years literally everyone who was anyone made their way to the great harvest festival celebrated on the plain of Athens.

The French historian of religion Le Clerc de Septchenes, writing toward the close of the eighteenth century, put it this way:

> According to Cicero, people came from all quarters to be initiated here. "Is there a single Greek, says Aristides, a

> single Barbarian so ignorant, so impious, as not to consider Eleusis as the common temple of the world?" That temple was built at a town in the neighborhood of Athens, on the ground that had first yielded the bounties of Ceres. It was remarkable for the magnificence of its architecture, as well as for its immense extent; and Strabo observes, that it would contain as many people as the largest Amphitheater.[13]

The power of the Eleusinian Mysteries lay in the fact that they possessed no dogma but, rather, involved certain sacred acts that engendered religious feeling and into which each successive age could project the symbolism it desired. Orthodox scholars, themselves unfamiliar with the reality-transforming power of plant hallucinogens, have fallen victim to the prejudiced attitude toward ecstasy that typifies the constipated patriarchal academy and hence have been baffled by the Mystery. And their bafflement has produced some of the most tortured of speculations.

> Albrecht Dieterich presumed that the object taken from the chest and in some way manipulated by the *mystes* was a phallus. This, however, met with the objection that Demeter was after all a female deity. Alfred Korte was therefore much applauded when he announced that it must be a female sexual symbol. Now everything seemed clear as day. By touching the "womb," as the sexual symbol was called, the *mystes* was reborn; and since such an act must after all have constituted the climax of the mysteries, Ludwig Noack went so far as to assume that the hierophant displayed this "womb" to the congregation in a blaze of light and that, beholding it, the initiates could no longer doubt their beatific lot as children of the goddess. It is difficult to report such notions without a smile.[14]

Indeed. Displaying a representation of the vagina might have riveted a room full of male Victorian classicists, but one would like to believe that the mystical wellspring of the classical world was something more than a peep show.

A PSYCHEDELIC MYSTERY?

There is little doubt that at Eleusis something was drunk by each initiate and each saw something during the initiation that was utterly unexpected, transformative, and capable of remaining with each participant as a powerful memory for the rest of their life. It is an incredible testament to the obtuseness of the scholars of the dominator society that not until 1964 did someone make bold to suggest that a hallucinogenic plant must have been involved. That person was the English poet Robert Graves in his essay "The Two Births of Dionysus":

> The secret which Demeter sent around the world from Eleusis in charge of her protégé Triptolemus is said to have been the art of sowing and harvesting grain. . . . Something is wrong here. Triptolemus belongs to the late second millennium B.C.; and grain, we now know, had been cultivated at Jericho and elsewhere since around 7000 B.C. So Triptolemus's news would have been no news. . . . Triptolemus's secret seems therefore concerned with hallucinogenic mushrooms, and my guess is that the priesthood at Eleusis had discovered an alternative hallucinogenic mushroom easier to handle than the *Amanita muscaria*; one that could be baked in sacrificial cakes, shaped like pigs or *phalloi*, without losing its hallucinogenic powers.[15]

This was the first of many observations Graves made on the underground tradition of mushroom use in prehistory. He suggested to the Wassons that they visit Mazatecan Mexico for evidence supporting their theories on the impact of intoxicating mushrooms on culture. Graves believed that recipes in classical sources for the preparation of the ritual Eleusinian beverage contained ingredients whose first letters could be arranged to spell out the word "mushroom"—the secret ingredient. Such a cypher is called an *ogham* after the similar poetic device in use in Irish riddlery and poetics. Graves readily grants that "you are at liberty to call me crazy," but then goes on to defend his thesis very well.

Perhaps we shall never know the nature of the hallucinogenic plants that lie behind the Eleusinian Mystery or that propelled the celebrants of Dionysus into a frenzy that was overwhelming to experience and frightening to behold. Graves, having opened the way to speculation on the botanical reality behind the Eleusinian sacrament, then had the pleasure of seeing his friend Wasson stride down that newly opened avenue of thought with a bold and convincing theory.

THE ERGOTIZED BEER THEORY

Wasson's notion, worked out in collaboration with his fellow sleuths Albert Hofmann and Carl Ruck and unveiled at a mushroom conference in San Francisco in 1977, was that Eleusis was a rite of visionary intoxication but mushrooms were not directly involved. Wasson gave cogency to much that had previously been obscure by arguing that the source of intoxication was an ergotized beer brewed from a strain of ergot fungus.

Some background is necessary to appreciate the neatness of this suggestion. Grain was somehow very important to the cult at Eleusis. The festival of the Mysteries was a harvest festival, as well as the celebration of a great agricultural secret and a mystery of the Mother Goddess and Dionysus. *Claviceps purpurea*, a small fungus that infects edible grains, produces ergot, a source of powerful alkaloids capable of causing hallucination (as well as of triggering the onset of labor and having a strong vasoconstrictive effect). The purple traditionally associated with the robe of Demeter may signify the distinctive purple color of the *sclerotia*, the ergot of commerce which are purple and are an asexual resting stage in the life cycle of the organism. Mycelium sprouts from them and aggregates to form the spore-containing *asci* which do look like tiny mushrooms, but they are not purple but rather light bluish.

Arguing for their theory, Wasson and his colleagues wrote:

> Clearly ergot of barley is the likely psychotropic ingredient in the Eleusinian potion. Its seeming symbiotic relationship to the barley signified an appropriate expropriation and transmutation of the Dionysian spirit to which the grain, Demeter's daughter, was lost in the nuptial embrace with earth. Grain and ergot together, moreover, were

> joined in a bisexual union as siblings, bearing at the time of the maiden's loss already the potential for her own return and for the birth of the phalloid son [the mushroom] that would grow from her body. A similar hermaphroditism occurs in the mythical traditions about the grotesquely fertile woman whose obscene jests were said to have cheered Demeter from her grief just before she drank the potion.[16]

Wasson and Hofmann's theory is bold and well argued. Certainly, their discussion of the scandal of 415 B.C., in which the Athenian noble Alcibiades was fined for having the Eleusinian sacrament in his home and using it for the entertainment of his friends, makes clear to even the most resistant skeptic that whatever the catalyst for ecstasy at Eleusis was, it was tangible.

The notion that Eleusinian rites were celebrated with ergotized beer is entirely consistent with the notion that they had historical roots in Minoan Crete. In 1900 Sir Arthur Evans, excavating near the palace of Knossos, unearthed vessels adorned with ears of barley in relief. He therefore assumed that a kind of beer had preceded wine on Crete. Kerényi believes that the small size of these vessels indicates they were used for a special kind of barley drink—the visionary sacrament of the Eleusinian mysteries—in rites "allegedly performed without secrecy at Knossos."[17]

Of course, "the burden of proof is on those who assert," and so far as I am aware no one has subjected the Wasson-Hofmann theory to the acid test. That would mean the actual brewing of a superior hallucinogen from a cereal grain infected with some strain of ergot. Until this is done the theory remains nothing more than well-argued speculation. One problem in particular needs to be dealt with: in documented instances in which large numbers of people have eaten ergot-infected grain, the result has been far from happy. Ergot is toxic. In A.D. 994, an outbreak of ergotism associated with infected grain killed nearly 40,000 people in France; an outbreak in 1129 killed about 1,200 people. Recently the historian Mary Kilbourne Matossian has argued that La Grande Peur of 1789, a peasant uprising pivotal in the French Revolution, had its roots in ergot-infected rye bread that constituted the bulk of the diet of the rural peasantry of the period. It has also been proposed that ergot-infected flour was a factor in the decline of the Roman Empire and in the

Salem witch burnings.[18] The following summarizes the apparent effects of ergotism:

> Two clinical types of ergotism have been described, the gangrenous and the convulsive. Gangrenous ergotism started with tingling in the fingers, then vomiting and diarrhea, followed within a few days by gangrene in the toes and fingers. Entire limbs were affected by a dry gangrene of the entire limb, followed by its separation. The convulsive form started the same way but was followed by painful spasms of the limb muscles which culminated in epileptic-like convulsions. Many patients become delirious.[19]

Clearly unpleasant experiences may lie ahead for those who set out to prove by self-experiment the Wasson-Hofmann theory concerning Eleusis. There are old mycologists, and there are bold mycologists, but there are no old, bold mycologists. As with Wasson's theory of the identity of Soma, the problem is to obtain a reliable intoxication from the assumed source of the intoxicant. If the source of the Eleusinian Mystery was ergotized beer, how could it have been taken for so many centuries without unpleasant side effects becoming a part of the legend?

There may be a way around these difficulties. *Claviceps paspali*, which preferentially infects barley instead of rye, may have a higher proportion of the psychoactive but less toxic "simple" ergot alkaloids (similar to those in morning glories) and a lower proportion of the toxic peptide-containing ergot alkaloids. Also, as Wasson and Hofmann reported in *The Road to Eleusis*, macerating the ergotized grain in water would effectively separate the water-soluble psychoactive alkaloids from the fat- or lipo-soluble toxic alkaloids.

GRAVES'S PSILOCYBIN THEORY

Should future research indicate that ergot played no part at Eleusis, then Graves's insistence that psilocybin mushrooms constituted the mystery would need to be looked at very carefully. Perhaps knowledge of the Ur-plant of the Goddess, *Stropharia cubensis*, or some other psilocybin-containing mushroom did survive, not only into

Minoan-Mycenaean times but even up until the final destruction of Eleusis.

Whatever its nature, the Eleusinian sacrament commanded the greatest respect and even love of the classical writers who invoked it: "Happy is he who, having seen these rites, goes below the hollow earth; for he knows the end of life and he knows its god-sent beginning," wrote the Greek poet Pindar. With the passing of Eleusis, the great broad river of partnership, Goddess worship, and hallucinogenic ecstasy that had flowed for over ten thousand years sank at last into that chthonic realm reserved for forgotten religions. Christianity's triumph ended the glorification of nature and planet as supreme spiritual forces. What Eisler called the "triumph of the blade" of dominator social models of paternalism and patriarchy was everywhere complete. Only a dim echo of the old ways continued to reverberate in the form of such underground concerns as alchemy, hermeticism, midwifery, and herbalism.

A HISTORICAL WATERSHED

With the eclipse of Minoan Crete and its Mysteries, humankind crossed a watershed into the progressively more vacant, more ego-dominated world, whose energies were coalescing into monotheism, patriarchy, and male domination. Henceforward the great society-shaping plant relationships of the Old World's past would decline to the status of "mysteries," esoteric pursuits of monied travelers and the religiously obsessed, and, later, cynical intelligence operatives.

As the Mysteries faded, the phonetic alphabet helped move consciousness toward a world emphasizing spoken and written language and away from the world of a gestalt pictographic awareness. These developments reinforced the emergence of the antivisionary dominator style of culture. The dark night of the planetary soul that we call Western civilization began.

9

ALCOHOL AND THE ALCHEMY OF SPIRIT

The ecstatic and orgiastic, visionary and boundary-dissolving experiences, the central mysteries of the mushroom religion, were the very factors in the human situation acting to keep our ancestors human. The commonality of feeling generated by the mushroom held the community together. The divine, inspiring power of the mushroom spoke through the bards and singers. The indwelling spirit of the mushroom moved the hand that carved bone and painted stone. Such things were a commonplace of the Edenic world of the Goddess. Life was lived not as we have chosen to imagine it, on the edge of mute bestiality, but rather, close to a dimension of spontaneous magical and linguistic expression that now shines only briefly in each of us at the pinnacle of experimental intoxication but that then was the empowered and enveloping reality: the presence of the Great Goddess.

NOSTALGIA FOR PARADISE

History is the story of our unfocused agony over the loss of this perfect human world, and then of our forgetting it altogether, denying it and in so doing, denying a part of ourselves. It is a story of relationships, quasi-symbiotic compacts, with plants that were made and broken. The consequence of not seeing ourselves as a part of the green engine of vegetable nature is the alienation and despair that surrounds us and threatens to make the future unbearable.

It took many centuries for the flame of Eleusis to gutter into extinction, for the partnership, Mother Goddess view of community and society to fade. Then came many centuries more of nostalgia for paradise and its rivers of heavenly Soma, a nostalgia that took new and varied forms as humans sought to satisfy the innate yearning for intoxication.

> All the natural narcotics, stimulants, relaxants and hallucinogens known to the modern botanist and pharmacologist were discovered by primitive man and have been in use from time immemorial. One of the first things that *Homo sapiens* did with his newly developed rationality and self-consciousness was to set them to work to find out a way to by-pass analytical thinking and to transcend or, in extreme cases, temporarily obliterate, the isolating awareness of the self. Trying all things that grew in field or forest, they held fast to that which, in this context, seemed good—everything, that is to say, that would change the quality of consciousness, would make it different, no matter how, from everyday feeling, perceiving and thinking.[1]

Over the next few chapters we will examine these substitutes for the original mushroom intoxicant of prehistory. Unfortunately, our survey will only serve to underscore how far we have fallen from the original dynamic equilibrium of the partnership paradise.

ALCOHOL AND HONEY

The great plant-drug complex that spans this cultural divide is alcohol. Alcohol has its roots in the deepest stratum of Archaic cultural activities. Ancient civilizations of the Near East were preoccupied with beer making; very early in the development of human culture,

FIGURE 17. The bee-headed dancing goddesses. From a gold ring found at Isopata near Knossos. The heads and hands are those of an insect. From Marija Gimbutas's *The Goddesses and Gods of Old Europe*, 1982, Figure 146, p. 185.

if not long before, the intoxicating effects of fermented honey and fruit juices must have been noticed.

Honey is a magical substance—a medicinal substance in all traditional cultures. As we have seen, it has been used to preserve both human bodies and mushrooms. Mead, or fermented honey, seems to have been the recreational drug of the Indo-European tribes. This was a cultural trait they shared with the mushroom-using pastoralists of the ancient Near East. One of the most astonishing murals unearthed at Çatal Hüyük apparently depicts the life cycle and metamorphosis of honeybees. (See Figure 9.)

The belief widely held in the classical world that bees were generated from the carcasses of cattle makes more sense if seen as an effort to connect bees as a source of honey and mead, the supplanting intoxicant, with cattle and the older mushroom cult. It may be that mead cults and mushroom cults that used honey as a preservative developed in close association with each other.

Honey is closely connected to the Great Goddess rites of the

Archaic Minoan civilization and is a prominent motif in the myths surrounding Dionysus (Figure 17). Dionysus was said by the Roman poet Ovid to have invented honey;[2] and the sacred ground on which the maenads, his handmaidens, performed their ritual dance was said to have flowed with milk, wine, and the "Nectre of bees." It was also said that honey dripped from the thyrsos staffs that the maenads carried. Kerényi, speaking of the honey offerings in Minoan religion, observes: "The honey offering given to the 'mistress of the labyrinth' carries the style of a much earlier period: that stage in which Minoan culture was still in contact with an 'age of honey.'"[3]

Each intoxicant, each effort to recapture the symbiotic balance of the human-mushroom relationship in the lost African Eden, is a paler, more distorted image of the original Mystery than the last. The devolution of sacramental elements in the religion of the ancient Near East must have led from mushrooms through fermented honey and juices to the emergence of the grape as the favored wine plant. Over time and often within the same cultures, fermented cereals and grains were manipulated experimentally to produce early types of beer.

WINE AND WOMAN

> Fruits rich in seed, such as pomegranates and figs, appear from the earliest times as symbols of fecundity. The vine and its juice has a long history of religious significance. Deified, like the Zoroastrian *haoma* and Vedic *soma*, its powers of exhilaration and intoxication were thought to be manifestations of divine possession. In the group of sacraments or "mysteries" that we shall examine, . . . the vine symbolizes especially the fruitfulness of woman, and its juice, mostly unfermented, is drunk ceremonially in order to promote the fertility of the womb.[4]

Wine played a central part in later Greek culture, so much so that in classical times the disturbing figure of ecstatic Dionysus was converted into the hairy-footed and lascivious wine-god Bacchus, the lord of orgy and, now, drunken revelry carried on in the traditional dominator style. The fermentation of grains and fruits must have been generally known and can claim no discoverer or point of origin.

Greek wines have always been somewhat puzzling to scholars. Their alcohol content could not have exceeded 14 percent since, when a fermentation process reaches this concentration, further formation of alcohol is inhibited. Yet Greek wines are sometimes described as requiring many dilutions before they could be drunk with comfort. This seems to suggest that Greek wines were more akin to extracts and tinctures of other plant essences than they were to wine as we know it today. This would have made them more chemically complex and therefore more intoxicating. The practice of adding resin to wine in Greece to make *retsina* may well hark back to times when other plants, perhaps belladonna or *Datura*, also went into wine.

Alcohol is the first example of a disturbing phenomenon that we will meet again and again in our discussion of differences in ancient and modern approaches to drug use and drug technology. Human use of alcohol in the form of fermented grains, juices, and mead is extremely ancient. Distilled spirits, in contrast, were not known to the ancients (though Pliny mentions one Roman wine so powerful that it burned when poured onto a fire). And today it is distilled alcohol that is the chief culprit among the drugs labeled "legal" and "recreational."

NATURAL AND SYNTHETIC DRUGS

Discussion of alcohol gives us our first opportunity to examine the distinction between natural and synthetic drugs, for though distilled alcohol waited for hundreds of years to be joined by a second example of a chemically refined intoxicant, it was the first highly concentrated and purified drug, the first synthetic drug. This distinction is very important for the argument to be made here. Alcoholism as a social and community problem appears to have been rare before the discovery of distillation. Just as heroin addiction was the malignant flower that sprang from the relatively benign habit of opium use, so distilled alcohol changed the sacred art of the brewer and the vintner into a profane economic engine for the consumption of human hopes.

It is no accident that alcohol was the first intoxicant to undergo this transformation. Alcohol can be fermented out of many kinds of fruits, grains, and plants, and so has been more widely experimented with than obscure and localized sources of intoxication. Indeed, fermentation is a natural process that in many cases is difficult to avoid. And fermented alcohol can be produced in prodigious—and hence commercial—amounts. The toddy palms of Southeast Asia produce debatably drinkable alcohol straight from the tree. Birds, raccoons, horses, and even wasps and butterflies are aware of the fleeting virtues that attend eating fermented fruit:

> In wild habitats most intoxications occur with the ingestion of fermented fruits, grains, or saps. Field teams have investigated dozens of cases, from Sumatra to the Sudan, involving creatures from bumblebees to bull elephants. The results? In natural habitats, most animals seek alcohol-laden food for the smells, tastes, calories, or nutrients they provide. The intoxications are side effects but not serious enough to deter future use.
>
> One sort of accidental intoxication occurs when tree sap is exposed to the proper temperature and ferments. The North American sapsuckers, a type of woodpecker, drill pit-like holes in trees that then fill with sap. The birds feed on the sap and insects attracted to the sap pits. They move on to other trees, literally "leaving the doors open" for the sap to ferment and intoxicate other animals before the tree heals over. The drinking of fermented sap has been held responsible for an array of abnormal behaviors observed in hummingbirds, squirrels, and unsuspecting sapsuckers.[5]

Alcohol can be distilled by using heat to vaporize it and separate it from its source, unlike alkaloids and indoles, which must somehow be extracted using solvents and then concentrated. This fact —that a simple water-cooled condenser can capture the vapor of alcohol and return it to liquid form—made it possible for alcohol to be the first intoxicant to be chemically "isolated." (The quality of being recaptured from its vaporous state is what gave rise to the practice of referring to distilled alcohol as "spirits.")

The first reference that we have to what might be a distilled form of alcohol occurs in the fourth century A.D. writings of the Chinese alchemist Ko Hung. In discussing recipes for the preparation of cinnabar, Ko Hung comments: "They are like wine that has been fermented once; it cannot be compared with the pure, clear wine that has been fermented nine times."[6] This statement seems to imply the knowledge of methods for the preparation of very strong clear alcohols, perhaps by the capture of alcohol vapor in wool from which could be wrung a relatively pure liquid alcohol.

ALCHEMY AND ALCOHOL

In the West the discovery of distilled alcohol is alternately credited to the alchemist Raymond Lully, about whom very little is known with certainty, or to his peer and companion in alchemical exploits, Arnoldus de Villanova. Lully's search for the true elixir led him to the preparation of *aqua vini*, the first brandy. According to Matheson, Lully was so awed by the wonders of *aqua vini* that he thought its discovery must surely herald the end of the world.[7] True to his alchemical roots, Lully made his universal panacea by fermenting wine in a double boiler of horse dung for twenty days before distilling it with a crude cold-water condenser. (See Figure 18). Lully did not hide his discovery; on the contrary, he invited others to make the elixir for themselves and hailed the product offered by Villanova as comparable to his own. Of alcohol he wrote, "The taste of it exceedeth all other tastes and the smell all other smells." It is, he says, "of marvelous use and commodity a little before the joining in battle to encourage the soldiers' minds."[8]

These discoveries of the intoxicating chemical agent lying behind the fermentation of fruit juices, honey, and grains were made, both in China and in Europe, by alchemists. Alchemy was a slowly evolving, loosely knit, and not mutually exclusive group of Gnostic and Hermetic theories concerning human origins and the dichotomy of spirit and matter. Its roots reached back deep into time, to at least Dynastic Egypt and the slow accumulation of jealously guarded secrets of processes for dyeing fabric, gilding metals, and mummifying bodies.

Upon those ancient foundations had risen an edifice of pre-Socratic, Pythagorean, and Hermetic philosophical ideas, which

FIGURE 18. Protochemical procedures and naive fantasy mingle in an alchemical process from the *Mutus Liber.* Courtesy of Fitz Hugh Ludlow Library.

ultimately came to revolve around the notion of the alchemical work as the task of somehow gathering into a unity and thereby rescuing the Divine Light that had been scattered through an alien and unfriendly universe by the fall of Adam. The natural world had come to be seen, by late Roman times, as a demonic and imprisoning shell. This was the spiritual legacy of the destruction of the partnership model of self and society and its replacement with the dominator model. The nostalgia for the Gaian Earth Mother was suppressed but could not, cannot, be ignored. Hence it reemerged in time in a clandestine form—as the alchemical theme of the *magma mater*, the mysterious mother matrix of the world, somehow everywhere, invisible yet potentially condensable into a visible manifestation of the universal panacea residing in nature.

In such an atmosphere of feverish and ontologically naive speculation, alchemy was able to thrive. Categories concerning self and matter, subject and object, were not yet fixed by the conventions introduced by phonetic alphabet and later exaggerated by print. It was not entirely clear to the alchemical investigators what about their labors was fancy, fact, or expectation.

It is ironic that this was the context for the discovery of a powerful mind-altering drug; that the spirit in alcohol, sensed and enjoyed in beer and wine brewed through the ages, became in the alchemical laboratories a demon, an elemental and fiery quintessence. And like those other quintessences that would follow it into existence, morphine and cocaine, the quintessence of the grape once passed through the furnace and the retorts of the alchemist had become deprived of its natural soul. That absence made it no longer a carrier of the vitality of the earth, no longer an echo of the lost paradise of prehistory, but rather something raw, untamed, and ultimately set against the human grain.

ALCOHOL AS SCOURGE

No other drug has had such a prolonged detrimental effect on human beings. The struggle to produce, control, and tax alcohol and to absorb its social consequences is a significant part of the story of the evolution of the mercantile empires of the eighteenth and nineteenth centuries. Alcohol and slavery often went hand in hand across the economic landscape. In many cases alcohol literally was slavery as the triangular trade of slaves, sugar, and rum and other practices of European civilization spread over the earth, subjugating other cultures. Sugar and the alcohol that could be made from it became a European obsession that severely distorted the demographics of tropical regions. For example, in the Dutch East Indies, now Indonesia, colonial policy paid women to produce as many children as possible, in order to provide workers for the labor-intensive cultivation of sugar. The modern legacy of this policy is that Java, formerly the center of the Dutch East Indies, is today the most overpopulated large island in the world. Most of the sugar ended up as distilled alcohol, and what was not exported to Europe was consumed by the local population. A "besotted underclass" was a permanent fixture of mercantile society whether in the home countries or the colonies.

And what of the psychology of alcoholism and alcohol use? Is there a gestalt of alcohol, and if there is, then what are its characteristics? I have implied that alcohol is the dominator drug par excellence. Alcohol has the effect of being libidinally stimulating at moderate doses at the same time that the ego feels empowered and social boundaries are felt to lose some of their restraining power. Often these feelings are accompanied by a sense of verbal facility ordinarily out of reach. The difficulty with all of this is that research findings suggest these fleeting effects are usually followed by a narrowing of awareness, a diminishing of ability to respond to social cues, and an infantile regression into loss of sexual performance, loss of general motor control, and consequent loss of self-esteem.

Moderation in drinking seems the obvious course. Yet alcoholism is a major and unremitting problem throughout global society. I believe that the alcohol abuse syndrome is symptomatic of the state of disequilibrium and tension existing between men and women and between the individual and society. Alcoholism is a condition of ego obsession and inability to resist the drive toward immediate gratification. The social domain in which the repression of women and the feminine is most graphically and brutally realized is that of the drunken episode or lifestyle. The darkest expressions of the terror and the anxiety engendered by severance from the maternal matrix have traditionally been acted out there. Wife beating without alcohol is like a circus without lions.

ALCOHOL AND THE FEMININE

The suppression of the feminine has been associated with the use of alcohol since very early times. One manifestation was the restriction of alcohol use to men. According to Lewin, women in ancient Rome were not allowed to drink wine.[9]

When Egnatius Mecenius's wife drank wine from a barrel, he beat her to death. He was later acquitted. Pompiliu Faunus had his wife whipped to death because she had drunk his wine. And yet another Roman woman of the gentry was condemned to die of hunger merely because she had opened the cupboard wherein were kept the keys to the wine cellar.

Dominator style hatred of women, general sexual ambivalence and anxiety, and alcohol culture conspired to create the peculiarly neurotic approach to sexuality that characterizes European civilization. Gone are the boundary-dissolving hallucinogenic orgies that diminished the ego of the individual and reasserted the values of the extended family and the tribe.

The dominator response to the need to release sexual tension in an ambience of alcohol is the dance hall, the bordello, and the institutionalized expansion of a new underclass—that of the "fallen woman." The prostitute is a convenience for the dominator style, with its fear and disgust of women; alcohol and its social institutions create the social space in which this fascination and disgust can be acted out without responsibility.

This is a difficult subject to address. Alcohol is used by millions of people, both men and women, and I will make no friends by taking the position that alcohol culture is not politically correct. Yet how can we explain the legal toleration for alcohol, the most destructive of all intoxicants, and the almost frenzied efforts to repress nearly all other drugs? Could it not be that we are willing to pay the terrible toll that alcohol extracts because it is allowing us to continue the repressive dominator style that keeps us all infantile and irresponsible participants in a dominator world characterized by the marketing of ungratified sexual fantasy?

If you find this difficult to believe, then think about the extent to which images of sexual desirability in our society are associated with images of sophisticated use of alcohol. How many women have their first sexual experiences in an atmosphere of alcohol use that ensures that these crucial experiences take place entirely on dominator terms? The strongest argument for the legalization of any drug is that society has been able to survive the legalization of alcohol. If we can tolerate the legal use of alcohol, what drug cannot be absorbed in the structure of society?

We can almost see toleration of alcohol as the distinguishing feature of Western culture. This tolerance is related not only to a dominator approach to sexual politics but also to, for example, a reliance on sugar and red meat, which are complementary to an alcohol lifestyle. In spite of natural food fads and a rise in dietary awareness, the typical American adult diet continues to be one of sugar, meat, and alcohol. This "burn out diet" is neither healthy nor ecologically sound; it promotes heart disease, abuse of the land, and toxic addiction and intoxication. It exemplifies, in short, everything that is wrong with us, everything that we have been left with as a result of an unhindered millennium of practicing the tenets of dominator culture. We have achieved the triumphs of the dominator style—triumphs of high technology and scientific method—largely through a suppression of the more untidy, emotional, and "merely felt" aspects of our existence. Alcohol has always been there when we needed to call upon it to propel us further down this same path. Alcohol helps nerve a man for battle, helps nerve men and women for love, and keeps an authentic perspective on self and world forever at bay. It is unsettling to realize that the delicately maintained web of diplomatic agreements and treaties standing between us and nuclear Armageddon was fabricated in the atmosphere of misguided sentimentality and blustering bravado that is typical of alcoholic personalities everywhere.

10
THE BALLAD OF THE DREAMING WEAVERS: CANNABIS AND CULTURE

No plant has been a continuous part of the human family longer than the hemp plant. Hemp seeds and remains of ancient cordage have been found in the earliest strata of many Eurasian habitation sites. Cannabis, a native of the heartlands of Central Asia, was spread throughout the world by human agency. It was introduced into Africa at a very early date, and cold-adapted strains traveled with the early human beings who crossed the land bridge into the New World. Because of its pandemic range and environmental adaptability, cannabis has had a major impact on human social forms and cultural self-images. When the resin of the cannabis plant is collected together into black sticky balls, its effects are comparable to the power of a hallucinogen, providing that the material is eaten. This is the classic hashish.

The thousands of names by which cannabis is known in hundreds of languages are testament not only to its cultural history and ubiquity but also its power to move the language-making faculty of the poetic soul. *Kunubu* it is called in an Assyrian letter tentatively dated 685 B.C.; a hundred years later it is referred to as *kannapu*, the root of the Greek and Latin *cannabis*. It is *bang*, *beng*, and *bhnj*; it is *ganja*, *gangika*, and *ganga*. *Asa* to the Japanese is *dagga* to the Hottentots; it is also *keif* and *keef* and *kerp* and *ma*.

American slang alone contains a prodigious number of words for cannabis. Even before 1940, before it was a part of mainstream white culture, cannabis was known as *muggles*, *mooter*, *reefer*, *greefa*, *griffo*, *Mary Warner*, *Mary Weaver*, *Mary Jane*, *Indian hay*, *loco weed*, *love weed*, *joy smoke*, *giggle smoke*, *bambalacha*, *mohasky*, *mu*, and *moocah*. Such terms were the mantras of an experientially oriented underclass religion that worshiped a jolly green goddess.[1]

HASHISH

Hashish is several thousands of years old, although at what point human beings began to gather and concentrate cannabis resin in this way is not clear. Smoking of cannabis products, the most efficient and rapid way of obtaining their effects, reached Europe rather late. In fact, smoking itself was only introduced into Europe when Columbus returned with tobacco from his second trip to the New World.

This is rather remarkable: a major human behavior pattern was unknown in Europe until quite recently. One might make the observation that Europeans generally seemed resistant to the development of innovative strategies of drug use. For example, the enema, another means of administering strong plant extracts, was also developed in the New World, by Indians of the equatorial Amazonian forests to whom natural rubber was familiar. Its development allowed experimentation with plants whose effects or taste were objectionable when taken orally.

It is not possible to say with certainty when cannabis was first smoked or, indeed, whether smoking was once part of the cultural repertoire of Old World peoples and then forgotten, only to be reintroduced from the New World at the time of the Spanish Conquest. For while smoking was unknown to the Greeks and the Romans, it may have flourished in the Old World in prehistoric times. Archaeological digging at Non Nak Tha in Thailand has yielded, in graves dated 15,000 B.P., the remains of animal bones that appear to have had plant material repeatedly burned in their hollow centers. The favorite instrument for the smoking of cannabis in India even to this day is a *chelum*, a simple wooden, ceramic, or soapstone tube that is packed with hashish and tobacco. How long have chelums been used in India is a matter of debate, but there can be little doubt that the method is extremely effective.

THE SCYTHIANS

The Scythians, a nomadic central Asian barbarian group who entered eastern Europe around 700 B.C., are the people who brought the use of cannabis to the European world. Herodotus describes their novel method of self-intoxication, a kind of cannabis sweat lodge:

> They have a sort of hemp growing in this country [Scythia], very like flax, except in thickness and height; in this respect the hemp is far superior: it grows both spontaneously and from cultivation. . . . When, therefore, the Scythians have taken some seed of this hemp, they creep under the cloths [of the sweat lodge] and then put the seed on the red hot stones; but this being put on smokes, and produces such a steam, that no Grecian vapour-bath would surpass it. The Scythians, transported by the vapour, shout aloud.[2]

Elsewhere Herodotus comments on another, similar method:

> [The Scythians] have discovered other trees that produce fruit of a peculiar kind, which the inhabitants, when they meet together in companies, and have lit a fire, throw on the fire, as they sit round in a circle; and that by inhaling the fumes of the burning fruit that has been thrown on, they become intoxicated by the odor, just as the Greeks do by wine; and that the more fruit that is thrown on the more intoxicated they become, until they rise up to dance and betake themselves to singing.[3]

The passage from Herodotus makes clear that though the Scythians had discovered that inhaling the smoke of cannabis was the most effective way to enjoy it, nevertheless they were unable to make the creative leap to the invention of the pipe or chelum! The Greek herbalist and natural scientist Dioscorides also described cannabis, but until effective smoking practices were adopted, it made no inroads into European and American cultures.

INDIA AND CHINA

Chinese tradition holds that hemp cultivation began as early as the twenty-eighth century B.C., when the emperor Shen-Nung taught the cultivation of hemp for fiber. And around A.D. 220 the physician Hoa-tho evidently recommended hemp preparations in wine as an anaesthetic: "After a certain number of days or the end of a month the patient finds he has recovered without having experienced the slightest pain during the operation."[4]

Cannabis was used and regarded as a plant of great spiritual power for many centuries in India before it was first smoked. Opium, too, seems to have been used for many centuries before the effectiveness of smoking it was discovered. Awareness of hemp in India cannot be documented before 1000 B.C., but by that time it was known as a remedy and the names for it in use in the earliest Indian pharmacopoeias indicate that its activity as a euphoriant was clearly understood. General awareness of the properties of cannabis grew very slowly and cannot be assumed to be widespread until around the tenth century A.D., only shortly before the Islamic invasion of Hindu India. Cannabis had associations with the esoteric, hence secret, side of Muslim and Hindu religiosity. Esoteric spirituality, the yogic practices of saddhus, and the emphasis on the direct experience of the transcendent are all little more than aspects of the veneration of cannabis in India. J. Campbell Oman, a late-nineteenth-century observer of Indian folkways, wrote:

> It would be an interesting philosophical study to endeavor to trace the influence of these powerful narcotics on the minds and bodies of the itinerant monks who habitually use them. We may be sure that these hemp drugs, known since very early times in the East, are not irresponsible for some of its wild dreamings.[5]

Oman touches here on a very fruitful theme—the degree to which the style and way of life of an entire culture can be imbued with the attitudes and assumptions engendered by a particular psychoactive plant or drug. There is something to the notion that the architectural styles and design motifs of Mughal Delhi or tenth-century Isfahan are somehow derivative of or inspired by the visions of hashish. And there is something to the notion that alcohol channeled the development of social forms and cultural self-image in feudal Europe. Aesthetic assumptions and styles are indices of the level and kind of understanding and awareness that a society sanctions. Each plant relationship will tend to accentuate some concerns and diminish others.

Outpourings of style and esthetically managed personal display are usually anathema to the nuts-and-bolts mentality of dominator cultures. In dominator cultures without any living traditions of use of plants that dissolve social conditioning, such displays are usually felt to be the prerogative of women. Men who focus on such concerns are often assumed to be homosexuals—that is, they are not following the accepted canons of male behavior within the dominator model. The longer hair lengths for men seen with the rise of marijuana use in the United States in the 1960s were a textbook case of an influx of apparently feminine values accompanying the use of a boundary-dissolving plant. The hysterical reaction to such a minor adjustment in folkways revealed the insecurity and sense of danger felt by the male ego in the presence of any factor that might tend to restore the importance of partnership in human affairs.

In this context, it is interesting to note that cannabis occurs in both a male and a female form. And it is the identification, care, and propagation of the female of the species that is the total concern of the grower interested in the narcotic power of the plant. This is because the resin is the exclusive product of the female plant. Not only do males not produce a usable drug, but if the pollen from male plants reaches females, the females will begin to "set" seed and will cease their production of resin. It is thus a kind of happy coincidence that the subjective effects of ingesting cannabis and the care and attention needed to produce a good resin strain both conspire to accentuate values that are oriented toward honoring and preserving the feminine.

Of all the pandemic plant intoxicants inhabiting the earth, cannabis is second only to mushrooms in its promotion of the social values and sensory ratios that typified the original partnership societies. How else are we to explain the unrelenting persecution of cannabis use in the face of overwhelming evidence that, of all the intoxicants ever used, cannabis is among the most benign? Its social consequences are negligible compared with those of alcohol. Cannabis is anathema to the dominator culture because it deconditions or decouples users from accepted values. Because of its subliminally psychedelic effect, cannabis, when pursued as a lifestyle, places a person in intuitive contact with less goal-oriented and less competitive behavior patterns. For these reasons marijuana is unwelcome in the modern office environment, while a drug such as coffee, which reinforces the values of industrial culture, is both welcomed and encouraged. Cannabis use is correctly sensed as heretical and deeply disloyal to the values of male dominance and stratified hierarchy. Legalization of marijuana is thus a complex issue, since it involves legitimating a social factor that might ameliorate or even modify ego-dominant values.

Legalization and taxation of cannabis would provide a tax base that could help clean up the national deficit. Instead, we continue to hurl millions of dollars into marijuana eradication, a policy that creates suspicion and a permanent criminal class in communities that are otherwise among the most law abiding in the country.

As indicated, society's contempt for the cannabis user is a thinly disguised contempt for the values of community and the feminine. How else to explain the media's need to endlessly repudiate the psychedelic drug use and underground social experiments of the sixties? The fear that the flower children engendered in the establishment becomes understandable when analyzed in the light of the idea that what confronted the establishment was an outbreak of genderless partnership thinking based on a diminished sense of self-importance.

The Roman natural historian Pliny (A.D. 23–79) reproduces a fragment from Democratus concerning a plant called *thalassaegle* or *potamaugis* that many scholars consider a reference to cannabis:

> Taken in drink it produces a delirium, which presents to the fancy visions of a most extraordinary nature. The theangelis, he says, grows upon Mount Libanus in Syria, upon the chain of mountains called Dicte in Crete, and at Babylon and Susa in Persia. An infusion of it imparts powers of divination to the Magi. The gelotophyllis too, is a plant found in Bactriana, and on the banks of the Borysthenes. Taken internally with myrrh and wine all sorts of visionary forms present themselves, exciting the most immoderate laughter.[6]

Dioscorides, writing during the first century, gave an excellent description of cannabis and mentions its use in rope making and medicine, but he says nothing of its intoxicating properties. Because the climate favored the growth of hemp and Islam encouraged its use over alcohol, in the Near Eastern and Arab worlds, cannabis became the intoxicant of choice for many. This predilection for hashish and cannabis was already very old at the time of the Prophet, which explains why alcohol is explicitly forbidden to the faithful but hashish is a matter of theological disputation. By A.D. 950 use and abuse of hashish is widespread enough that it comes to occupy a prominent position in the literature of the period. A perfect encapsulation of the attitudes of dominator society toward cannabis is contained in the following, one of the earliest descriptions we possess of addictive behavior with this plant:

> A Moslem priest exhorting in the mosque against the use of "beng," a plant of which the principal quality is to intoxicate and induce sleep, was so carried away with the violence of his discourse that a paper containing some of the prohibited drug which often enslaved him fell from his breast into the midst of his audience. The priest without loss of countenance cried immediately, "There is this en-

> emy, this demon of which I have told you; the force of my words has put it to flight, take care that in quitting me it does not hurl itself on one of you and possess him." No one dared to touch it; after the sermon, the zealous sophist recovered his "beng."[7]

As this story makes clear, the ego of the monotheist is capable of the most extraordinary feats of self-delusion.

CANNABIS AND THE LANGUAGE OF STORY

Cannabis is a multipurpose plant: it very early came to the attention of hunter-gatherers as a source of cordage for weaving and rope making. But unlike other cordage plants—the flax of central Asia or the *chimbira* of the Amazon—cannabis is also psychoactive. In this context, it is interesting to note that the English vocabulary that refers to spoken discourse is often the same as that used to describe cordage-making and weaving. One weaves a story, or unravels an incident, or spins a yarn. We follow the thread of a story and stitch together an excuse. Lies are made from whole cloth, reality is an endless golden braid. Does this shared vocabulary reflect an ancient connection between the intoxicating hemp plant and the intellectual processes that lay behind the discovery of the art of weaving and of storytelling? I suggest that such may well be the case. Cannabis was the most likely plant candidate to replace the sacred psilocybin mushrooms of the older cultures of the Near East. Though this transition from mushrooms to cannabis lies far in the past, its legacy to the present era is the association of cannabis with the style of the partnership society. And, indeed, the growing presence of cannabis in Vedic society and later in Islam may have acted to slow the rise of dominator values. Certainly it gave encouragement to heterodox forces—Shivites in the case of Hinduism and Sufis in the case of Islam—who made no secret of their reliance on cannabis as a source of religious inspiration that was particularly feminine in emphasis.

The role of cannabis in European society is complex. Marco Polo, whose exploits and travel descriptions of the mysterious East did so much to enrich and catalyze the European imagination, gave one of the first and most widely read accounts of the use of hashish when he repeated the popular folktale of the "Old Man of the Mountain" Ibn el Sabah, reputed leader of the violent cult of the hashishin, the infamous sect of assassins. According to the legend, young men wishing to be initiated into the sect were given large doses of hashish and then introduced into an "artificial paradise"—a hidden valley of exotic floral gardens, splashing fountains, and nubile young women. They were told that return to this land of dreams was only possible after they had carried out certain acts of political murder. Indeed, "hashishin" and "assassin" are thought to be etymologically related. The truth of this old story is widely disputed, but there can be no doubt that it was the circulation of the story in Europe that gave cannabis its blackened reputation and its fascination.

Some five hundred years after Marco Polo, French administrators of Napoleonic Egypt failed utterly in their efforts to control the production and sale of cannabis preparations. In response to a ban on sales, Greek smugglers immediately began a lucrative underground business of importing hashish into Egypt.

Militarily, Napoleon's expedition into Egypt was a failure, but as an effort at the cross-fertilization of disparate cultures it was a resounding success. Napoleon took with him into Egypt an excellent library and 175 scholars who observed, sketched, and collected linguistic and cultural information. This effort ultimately resulted in the publication of twenty-four volumes *(Description d'Egypte)* between 1809 and 1813. These volumes inspired a wide variety of travel books and in general were a tremendous stimulus to the European imagination.

ORIENTOMANIA AND CANNABIS IN EUROPE

While Napoleon struggled with the prevalence of cannabis use in Egypt, new intellectual forces were stirring in Europe. Romanticism, Orientomania, and a fascination with psychology and the paranormal all combined with the well-established upper-class craze for opium and the opium tincture, laudanum, to create a climate in which the reputed pleasures of hashish could be explored by daring and unconventional souls. The legal and intellectual ambience of drug taking in the early nineteenth century could hardly have been more different from that of our own times. Opium and hashish were not controlled substances, and no opprobrium was attached to their use. Tobacco and coffee had long since been introduced into Europe and become indispensable parts of the rituals of European civilization, so it was not surprising that the extravagant tales of travelers concerning narcotic raptures and vistas of transcendental ecstasy acted to promote experimentation with cannabis.

By the early 1840s a group of French writers, among them Théophile Gautier, Baudelaire, Gérard de Nerval, Dumas, and Balzac, as well as a number of sculptors, painters, and other Bohemians, had formed the now famous "Club des Hachischins." The club held weekly meetings in damask-hung rooms of the Hôtel Luzan in Ile St.-Louis in Paris. At these meetings, world traveler and psychiatrist J. J. Moreau de Tours provided a form of jellied Algerian hashish called *dawamesc*. The meetings were the private explorations of successful and respected literary figures. Nevertheless, only a few years later, during the Paris uprising of 1848, the student firebrands carried banners through the streets demanding free availability of cannabis and ether.

In 1842, the English physician W. B. O'Shaughnessy became the first to introduce *ganja*, potent Indian hemp, to England, in his *Bengal Pharmacopeia*. Cannabis became a part of English medical practice and hence a part of the inventory of every English apothecary.

The relation between opium and hashish in the shaping of the European imagination is complex and synergistic. Opium has a much longer history of wide use in the West than has cannabis. Opium was known and used by physicians since at least late Egyptian and Minoan times, and it played a major part in the late, decadent phase of Minoan religion. Cannabis was introduced into Europe later and largely as a consequence of the interest in altered states that had already been kindled by opium enthusiasts.

Though cannabis had been used in the East for many centuries, it is very unlikely that more than a handful of Europeans were aware of its existence before the sensational account of Marco Polo appeared around 1290. In spite of the fact that the German physician

Johannus Weier mentioned the use of hashish by groups of witches in the sixteenth century, drugs based on hemp were absent from the materia medica of alchemy and were probably not brought into Europe in any quantity until O'Shaughnessy and his French contemporary, Aubert-Roche, advocated their use around 1840.

In 1845 J. J. Moreau de Tours published his *Du Hachisch et de l'Alienation Mentale (Hashish and Mental Illness)*. His detailed accounts of the effects of hashish sparked interest in both medical and literary circles, and set off a wave of experimentation. Even so, interest in hashish never traveled much beyond the Parisian circles in which Moreau himself moved. Hashish eating never became a European craze in the nineteenth century; use of hashish continued to be mostly confined to the Near and Middle East.

CANNABIS AND NINETEENTH-CENTURY AMERICA

It was not the English or the French but the Americans who created a literature around the charms and phantasmagoria of hashish. In so doing they were following the example of English opium habitués such as Coleridge and De Quincey. Thus their writings were strongly influenced by the "joys and horrors" style that had made De Quincey's name a household word. Their descriptions of the effects of cannabis make it abundantly clear that for them it had all the impact of a shattering metaphysical revelation. Today hashish eating, save for the occasional holiday cannabis cookie, is almost unknown as a method of cannabis ingestion; for us moderns, cannabis is inevitably something that one smokes. This was not true for the nineteenth century, which seems always to have eaten its hashish in the form of confections imported from the Middle East. These visions and the resulting intoxications leave no doubt that this method turns hashish into a powerful engine for the exploration of inner vistas of fantasy and awareness. The first exploratory journey into the teeming cosmos of cannabis to appear in print was an account by American traveler Bayard Taylor first published in *Atlantic Monthly* in 1854:

The sense of limitation—of the confinement of our senses within the bounds of our own flesh and blood—instantly fell away. The walls of my frame were burst outward and tumbled into ruin; and, without thinking what form I wore—losing sight even of all idea of form—I felt that I existed throughout a vast extent of space . . . the spirit (demon shall I rather say?) of Hasheesh had entire possession of me. I was cast upon the flood of his illusions, and drifted helplessly whithersoever they might choose to bear me. The thrills which ran through my nervous system became more rapid and fierce, accompanied with sensations that steeped my whole being in unutterable rapture. I was encompassed by a sea of light, through which played the pure, harmonious colors that are born of light. While endeavoring, in broken expression, to describe my feelings to my friends, who sat looking upon me incredulously—not yet having been affected by the drug—I suddenly found myself at the foot of the great Pyramid of Cheops. The tapering courses of yellow limestone gleamed like gold in the sun, and the pile rose so high that it seemed to lean for support upon the blue arch of the sky. I wished to ascend it, and the wish alone placed me immediately upon its apex, lifted thousands of feet above the wheat-fields and palm-groves of Egypt. I cast my eyes downward, and, to my astonishment, saw that it was built, not of limestone, but of huge square plugs of Cavendish tobacco! Words cannot paint the overwhelming sense of the ludicrous which I then experienced. I writhed on my chair in an agony of laughter, which was only relieved by the vision melting away like a dissolving view; till, out of my confusion of indistinct images and fragments of images, another and more wonderful vision arose.

The more vividly I recall the scene which followed, the more carefully I restore its different features, and separate the many threads of sensation which it wove into one gorgeous web, the more I despair of representing its exceeding glory. I was moving over the Desert, not upon the rocking dromedary, but seated in a barque made of mother-of-pearl, and studded with jewels of surpassing lustre. The sand was of

> grains of gold, and my keel slid through them without jar or sound. The air was radiant with excess of light, though no sun was to be seen. I inhaled the most delicious perfumes; and harmonies, such as Beethoven may have heard in dreams, but never wrote, floated around me. The atmosphere itself was light, odor, music; and each and all sublimated beyond anything the sober senses are capable of receiving. Before me—for a thousand leagues, as it seemed—stretched a vista of rainbows, whose colors gleamed with the splendor of gems—arches of living amethyst, sapphire, emerald, topaz, and ruby. By thousands and tens of thousands, they flew past me, as my dazzling barge sped down the magnificent arcade; yet the vista still stretched as far as ever before me. I revelled in a sensuous elysium, which was perfect, because no sense was left ungratified. But beyond all, my mind was filled with a boundless feeling of triumph.[8]

Such descriptions go a long way toward making clear why the "artificial paradise" was so alluring to the Romantic imagination: it was almost as though one were made for the other. And, indeed, the Romantics, with their attention to the dramatic moods of nature and their cultivation of a sensitivity that their critics found "feminine," bear all the signs of an incipient partnership revival. With the reportage of Bayard Taylor we are firmly in the domain of modern drug literature and modern values vis-à-vis the content of the intoxication. Taylor is impressed by the beauty, power, and *general depth of information* contained in the experience. His approach is not hedonistic but knowledge seeking, and for him as for us, the drug states raise questions about human psychology.

EVOLVING DRUG ATTITUDES

This "scientific" attitude was typical of the nineteenth-century literate user of opium and hashish. Usually investigators began their involvement with these substances in order to "fire the creative imagination" or for a vaguely defined "inspiration." Similar motives were behind the use of marijuana by the writers of the Beat Generation, as well as the jazz artists before them and the rockers after them. Few myths of the underground culture invite as much current scorn as the notion that cannabis could contribute to a creative lifestyle. Nevertheless, a portion of the cannabis-using community continues to use it in this way.

The pharmacological profile of a drug defines only some of its parameters; the context—or "setting," in Leary and Metzner's fortunate turn of phrase—is at least as important. The "recreational" context for substance use, as currently understood in the United States, is an atmosphere that trivializes the cognitive impact of the substance used. Low doses of most drugs that affect the central nervous system are felt by the organism as artificial stimulation or energy, which can be directed outward in the form of physical activity in order both to express the energy and to quench it. This pharmacological fact lies behind most of the recreational drug craze whether legal or illegal. An environment dense with social signals, noise, and visual distraction—a nightclub, for example—is typical of the culturally validated context for use of recreational drugs.

In our culture, private drug taking is viewed as dubious; solitary drug use is viewed as positively morbid; and, indeed, all introspection is seen this way. The Archaic model for use of psychoactive plants, including cannabis, is quite the opposite. Ritual, isolation, and sensory deprivation are the techniques used by the Archaic shaman seeking to journey in the world of the spirits and ancestors. There is no doubt that cannabis is trivialized as a commodity and is degraded by the designation "recreational drug," but there is also no doubt that when used occasionally in a context of ritual and culturally reinforced expectation of a transformation of consciousness, cannabis is capable of nearly the full spectrum of psychedelic effects associated with hallucinogens.

FITZ HUGH LUDLOW

After Bayard Taylor the next great commentator on the phenomenon of hashish was the irrepressible Fitz Hugh Ludlow. This little-known bon vivant of nineteenth-century literature began a tradition of pharmo-picaresque literature that would find later practitioners in William Burroughs and Hunter S. Thompson. Ludlow, as a freshman at Union College in 1855, decided to explore scientifically the powers of hashish while attending a student tea:

> I was sitting at the tea table when the thrill smote me. I had handed my cup to Miss M'Ilvaine to replenish for the first time, and as she was about restoring it to me brimming with the draught which cheers but not inebriates. I should be loath to calculate the arc through which her hand appeared to me to travel on its way to the side of my plate. The wall grew populous with dancing satyrs; Chinese mandarins nodded idiotically in all the corners, and I felt strongly the necessity of leaving the table before I betrayed myself.[9]

There is in Ludlow's cannabis reportage a wonderful distillation of all that was zany in the Yankee transcendentalist approach. Ludlow creates a literary persona not unlike the poet John Shade in Nabokov's *Pale Fire*, a character who allows us to see deeper into his predicament than he can see himself. Part genius, part madman, Ludlow lies halfway between Captain Ahab and P. T. Barnum, a kind of Mark Twain on hashish. There is a wonderful charm to his free-spirited, pseudoscientific openness as he makes his way into the shifting dunescapes of the world of hashish:

> How far hasheesh throws light upon the most interior of the mental arcana is a question which will be dogmatically decided in two diametrically opposite ways. The man who believes in nothing which does not, in some way, become tangent to his bodily organs will instinctively withdraw himself into the fortress of what he supposes to be antique common sense, and cry "madman!" from within. He will reject all of experience under stimulus, and the facts which it professedly evolved as truth, with the final and unanswerable verdict of insanity.
>
> There is another class of men which has its type in him who, while acknowledging the corporeal senses as very important in the present nutriment and muniment of our being, is convinced that they give him appearances alone; not things as they are in their essence and their law, classified harmoniously with reference to their source, but only as

they affect him through the different adits of the body. This man will be prone to believe the Mind, in its prerogative of the only self-conscious being in the universe, has the right and the capacity to turn inward to itself for an answer to the puzzling enigmas of the world. . . .

Arguing thus, the man, albeit a visionary, will recognize the possibility of discovering from mind, in some of its extraordinarily awakened states, a truth, or a collection of truths, which do not become manifest in his every day condition.[10]

CANNABIS IN THE TWENTIETH CENTURY

The history of cannabis in the United States after Ludlow was at first a happy one. Cannabis use was neither stigmatized nor popularized. This situation lasted until the early 1930s, when the crusades of Harry J. Anslinger, U.S. Commissioner of Narcotics, created a public hysteria. Anslinger appears to have acted largely at the behest of American chemical and petrochemical companies interested in eliminating hemp as a competitor in the areas of lubricants, food, plastics, and fiber.[11]

Anslinger and the yellow press characterized cannabis as the "weed of death." William Randolph Hearst popularized the term "marijuana" with a clear intent of linking it to a mistrusted dark-skinned underclass. Yet it has been extraordinarily difficult for science to state exactly what the objections to the cannabis habit are. Patterns of government funding for research make it virtually certain that "Caesar will hear only what is pleasing to Caesar."

Despite all the pressures brought against it, cannabis use rose until today cannabis may well be America's *single largest agricultural product*. This is one of the most persistent aspects of the great paradigm shift that I am here calling the Archaic Revival. It indicates that the innate drive to restore the psychological balance typifying the partnership society, once it finds a suitable vehicle, is not easily deterred. Everything about cannabis that makes it inimical to contemporary bourgeois values endears it to the Archaic Revival. It diminishes the power of ego, has a mitigating effect on competitiveness, causes one to question authority, and reinforces the notion of the merely relative importance of social values.

No other drug can compete with cannabis for its ability to satisfy the innate yearnings for Archaic boundary dissolution and yet leave intact the structures of ordinary society. If every alcoholic were a pothead, if every crack user were a pothead, if every smoker smoked only cannabis, the social consequences of the "drug problem" would be transformed. Yet, as a society we are not ready to discuss the possibility of self-managed addictions and the possibility of intelligently choosing the plants we ally ourselves to. In time, and perhaps out of desperation, this will come.

III

HELL

11

COMPLACENCIES OF THE PEIGNOIR: SUGAR, COFFEE, TEA, AND CHOCOLATE

Long ago, motivated by dwindling resources and changing climate, our protohominid ancestors learned to test the natural products in the environment as sources of food. Modern primates such as baboons still do this. An unusual or never before encountered source of food is approached gingerly, examined carefully for visual appearance and odor, and then tentatively placed in the mouth and held there, not swallowed. After a few moments the animal makes the decision to swallow the morsel or to spit it out. Such a procedure has been repeated uncounted times over the long ages of human dietary definition.

Obviously a balance must be struck between excluding foods that would be outright injurious to the individual's health and reproductive ability and including as many sources of nutrition as possible. Evolutionary logic dictates that in situations of food scarcity those animals able and willing to tolerate many marginal foods will be more evolutionarily successful than those that can accept only a limited number of items into their diet. In other words, there will be pressure on a given animal to broaden its definition of what are acceptable foods by broadening its tastes.

BROADENING OUR TASTE

Broadening of tastes or acquiring a taste is a process that is learned; it is a process with both a psychological and biochemical component. The process of acquiring a taste is an extremely complex one. On the one hand, it entails overcoming the inertia of established habits, those habits that exclude the potential new food item, seeing it as exotic, unfamiliar, poisonous, or associated with enemies or social outcasts. And on the other hand, it involves an adaptation to a chemically exotic food. This process brings into action involuntary systems such as the immune system; it also involves psychological mechanisms, such as wanting to accept the new food item for reasons that may be as much social as nutritional. In the case of hallucinogenic plants, the shifts in self-image and societal role that often follow their acceptance are rapid and massive. But we should remember that hallucinogens are at the dramatic end of this scale.

What of the countless plants that impart flavoring but confer little nutritional value and negligible psychoactivity? They, too, have managed to become items habitually used by human beings. In fact, they went from being the exotic luxuries of a tiny leisure class in Roman times to commercial commodities that focused the vast European efforts at exploration and colonization, that drove the engines of mercantilism and empire building which replaced the inward-turned medieval stasis in Christian Europe.

"Variety is the spice of life" is an adage familiar to us all. Yet when we examine the impact of plants and plant products on the history of human beings it seems more true to say that "Spice is the variety of life." Medieval times—and their close—are a case in point.

Dominator culture has never been more powerfully entrenched than in Christian Europe after the eclipse of the Roman Empire. And it probably is safe to say that scarcely ever have human populations existed in such a prolonged situation of drug scarcity and lack of chemical stimulation. Variety, which promotes learning and eases boredom, had too long been suppressed in Europe.

Medieval Europe was one of the most constipated, neurotic, and woman-hating societies ever to exist. It was a society dying to escape from itself, a society obsessed with moral rectitude and sexual repression.

It was a society chained to the land, ruled by gouty, beef-eating men wearing dresses but suppressing women. Is it any wonder, then, that dyes and spices, hardly the stuff of social revolutions, became an absolute mania for medieval Europe? Such was the strength of this mania that the arts of shipbuilding and navigation and the banking and the trading industries all turned to serve the near-addiction that most Europeans felt for these things. Spices gave food, and hence life, a variety never known before. Dyes, new dyeing techniques, and exotic fabrics revolutionized fashion.

LIFE WITHOUT SPICE

It is difficult for most people born into a society of abundance, sensual gratification, and high-definition TV to imagine the stultifying dullness of most of the societies of the past. The "splendor" of the great societies of the past was essentially just a display of variety—variety in colors, fabrics, materials, and visual design. Such displays of variety were particularly the prerogative of the ruler and the court. The novelty of the costumes and the appointments of the court was somehow a direct index of its power. Thus it was when the emerging bourgeoisie of the late Middle Ages began importing dyes and spices, silks and fine manufactured objects into Europe.

I can personally attest to the power of color and variety over the human imagination. My periods of jungle isolation doing fieldwork in the Upper Amazon taught me how quickly the bewildering multiplicity of civilized life can be forgotten and then hungered for almost like the withdrawal from a powerful drug. After weeks in the jungle, one's mind is filled with plans for the restaurants to be visited once back in civilization, the music to be heard, the movies seen. Once, after many days in the rain forest, I went to a village to ask permission to make plant collections in the tribal area. The only "high-tech" intrusion into the primitive circumstances of the tribe was a cheesecake calendar brought from Iquitos and proudly affixed to the thatched wall directly behind the headman of the village. As

I talked with him my gaze returned again and again to the calendar, not the content but the colors. Magenta, cyan, and apricot—the terrible and obsessive attraction to variety was as haunting as the lure of any drug!

The dyes and spices of the more technically advanced and esthetically refined world of Islam entered the bloodstream of dreary Christian Europe with the force of a hallucinogenic drug. Cinnamon, cloves, nutmeg, mace, and cardamom, and dozens of other exotic spices, flavorings, and dyes, arrived to brighten the palate and the wardrobe of a wool-swaddled, beer and bread culture. Our own culture during the last few years has seen a similar though more superficial trend in the rise of the yuppie craze for novelty and new exotic restaurants from ethnic to nouvelle.

As schoolchildren we are taught that the spice trade ended the Middle Ages and created the basis of modern trade and commerce; what we are not made aware of is the fact that the breakup of Christian medieval Europe occurred as a result of an epidemic obsession with the new, the exotic, and the delightful—in short with consciousness-expanding substances. Drugs such as coffee, wormwood, and opium, dyes, silks, rare woods, gems, and even human beings were brought back to Europe and displayed almost like the plunder of a looted extraterrestrial civilization. The notion of oriental splendor—with its luxury, its sensuality, and its unexpectedly outré design motifs—acted to transform not only esthetic conventions, but canons of social behavior and individual self-image. The names of cities of the silk road, such as Samarkand and Ecbatana, became mantras, bespeaking worlds of refinement and luxury previously associated only with paradise. Social boundaries were dissolved; old problems were seen in a new light; and new secular classes emerged to challenge the power monopoly of popes and kings.

In short, there was a sudden acceleration of novelty and appearance of new social forms, the telltale tracks of a quantum leap forward in the power of the European imagination. Once again, pursuit of plants and of the mental stimulation they induce was propelling a portion of the human family into experimentation with new social forms, new technologies, and a sudden expansion of language and imagination. Pressure to expand the spice trade remade the arts of navigation, shipbuilding, diplomacy, warfare, geography and economic planning. Once again, the unconscious drive to mimic and thus partly recapture the lost symbiosis with the vegetable world was acting as a catalyst to dietary experimentation and to a restless quest for new plants and new relationships with plants, including new forms of intoxication.

ENTER SUGAR

When the thirst for variety was slaked by massive and continuous importation of spices, dyes, and flavorings, the infrastructure that had been put in place turned its attention to fulfilling other cravings for variety—specifically, to the production and shipping of sugar, of chocolate, tea, and coffee, and of distilled alcohol, all of which are drugs. Our present global trading system was created to cater to people's inherent need for variety and stimulation. It did this with a single-minded intensity that brooked no interference from the church or the state. Neither moral scruples nor physical barriers were able to stand in its way. Now we may appear to ourselves to have succeeded only too well—now any "spice" or drug, no matter how restricted its traditional area of use, can be identified and produced or synthesized for rapid export and sale to hungry markets anywhere on the globe.

Worldwide pandemics of substance abuse now become possible. The importation of tobacco smoking into Europe in the sixteenth century was the first and most obvious example. It was followed by many others, ranging from the forced spread of opium use in China by the British through the opium craze in eighteenth-century England to the spread of distilled alcohol abuse among the North American Indian tribes.

Of the many new commodities that made their way into Europe during the breakup of the medieval stasis, one in particular emerged as the new spice or drug of choice. This was cane sugar. Sugar had been known for centuries as a rare medicinal substance. The Romans knew that it was derived from a bamboolike grass. But the tropical conditions needed for the cultivation of sugarcane ensured that sugar would be a rare and imported commodity in Europe. Only in the nineteenth century, at the encouragement of Napoleon I, were sugar beets developed as an alternative to cane sugar.

Sugarcane is known to occur as a wild plant, and the genus is well represented in tropical Asia and at least five species are native to India. Sugarcane, *Saccharum officinarum*, has doubtless undergone considerable hybridization during its long history of domestication. The Persian king Khusraw I (A.D. 531–578), whose court was near Jundi-Shapur, dispatched envoys to India to investigate rumors of exotic drugs:

> Among those [drugs] brought to Jundi-Shapur from India was *sukkar* (Persian *shakar* or *shakkar*, Sanskrit *sarkara*), our sugar, unknown to Herodotus and Ktesias, but known to Nearchus and Onesicritus as "reed Honey," supposed to have been made from reeds by bees. Legend relates that Khusraw discovered a store of sugar amongst the treasures taken in 527 at the capture of Dastigrid. The juice of the sugar cane was purified and made into sugar in India about A.D. 300, and now the cane began to be cultivated about Jundi-Shapur, where there were sugar mills at an early date. At that time and for long afterwards sugar was used only to sweeten otherwise bitter medicines, it was not until much later that it began to replace honey as an ordinary means of sweetening.[1]

Sugar reached England around 1319 and was popular in Sweden by 1390. It was an expensive and exotic novelty, mostly found in its traditional role in medicine: sugar made palatable the foul-tasting mixture of medicinal herbs, entrails, and other materials typical of the medieval pharmacopeia. In the age before antibiotics, it was commonly used to pack wounds before binding them, as the desiccant action of the sugar may have aided healing.

The Spanish planted sugarcane in their Caribbean holdings, and they can claim the dubious distinction of introducing slavery into the New World for the purpose of producing sugar:

> Until 1550 the only sugar imported from the Western hemisphere consisted of a few loaves brought as proof of the possibility of production, or as mere curiosities. The plantings in West Atlantic islands and the New World had no effect on production, distribution, or prices until the

latter half of the sixteenth century, and only became dominant from about 1650.[2]

SUGAR AS ADDICTION

Is it stretching a point to discuss sugar in a history of human drug use? It is not. Sugar abuse is the world's least discussed and most widespread addiction. And it is one of the hardest of all habits to kick. Sugar addicts may be maintenance users or they may be binge eaters. The depths of serious sugar addiction are exemplified by bulimics who may binge on sugar-saturated food and then induce vomiting or use a laxative purge to enable them to eat more sugar. Imagine if a similar practice were associated with heroin addiction—how much more odious and insidious the use of heroin would then seem! As with all stimulants, ingestion of sugar is followed by a brief euphoric "rush," which is itself followed by depression and guilt. Sugar addiction rarely occurs alone as a syndrome; mixed addictions—for example, sugar and caffeine—are more common.

There are other destructive patterns of drug use that accompany sugar abuse. Some addicts use diet pills to help them control their soaring body weight, and then tranquilizers to mitigate the jitteriness caused by the diet pills. Sugar abuse is often involved in the development of serious alcohol abuse; an absolute correlation has been shown between high sugar consumption and high alcohol intake outside meals. After alcohol and tobacco, sugar is the most damaging addictive substance consumed by human beings. Its uncontrolled use can be a major chemical dependence.

In describing sugar addicts, Janice K. Phelps has said:

> The people we are describing are addictive people who are indeed addicted to one of the most powerful substances to be found anywhere—the refined sugars. Their addiction to sugar is a real, harmful, highly damaging health problem, just as debilitating as addiction to any other substance. Like any addiction, when their chemical isn't supplied, they suffer identifiable withdrawal symptoms; like any addiction, the process of feeding their physiological hunger with a chemical is destructive to the body; and like any addiction,

the point may be reached when supplying the chemical becomes as painful as withdrawing from it. The cycle of chemical dependence becomes both entrenched and intolerable.[3]

SUGAR AND SLAVERY

The distortion and dehumanizing of human institutions and human lives caused by crack cocaine today is nothing compared with what the European desire for sugar did in the seventeenth and eighteenth centuries. One may argue that something approaching slave labor is typical of the early stages with cocaine production but the difference is that it is not slavery sanctioned by mendacious popes and openly pursued by corrupt but legitimate governments. A further difference must be noted: brutal as it is, the modern drug trade is not involved in anything resembling the wholesale kidnapping, transporting, and mass murder of huge populations as was done to further the process of sugar production.

True, the roots of slavery in Europe reach far back. During the golden age of Periclean Athens fully two-thirds of the city's residents were slaves; in Italy at the time of Julius Caesar, perhaps one-half of the population were slaves. Under the Roman Imperium slavery became increasingly insupportable: slaves had no civil rights and in court disputes their testimony was acceptable only if it had been obtained by torture. If a slaveholder were to die suddenly or under suspicious circumstances, then all of his slaves, without regard to guilt or innocence, were quickly put to death. It is fair to say that the reliance of the Imperium on the institution of slavery must mitigate any admiration that we might feel for the "grandeur that was Rome." In truth, the grandeur of Rome was the grandeur of a pig sty masquerading as a military brothel.

Slavery diminished with the dissolution of the empire, as all social institutions dissolved into the chaos of the early Dark Ages. Feudalism replaced slavery with serfdom. Serfdom was somewhat better than slavery: a serf could at least maintain a home, marry, till the land, and participate in communal life. Most important, perhaps, a serf could not be separated from or transported off the land. When the land was sold the serf nearly always went with it.

In 1432 Prince Henry the Navigator of Portugal, who was more manager and entrepreneur than explorer, established the first commercial cane sugar plantation in Madeira. Plantings of sugar were made in the other eastern Atlantic holdings of Portugal more than sixty years before there was contact with the New World. More than a thousand men—including debtors, convicts, and unconverted Jews—were taken from Europe to work in the sugar operations. Their condition was one of quasi-servitude—somewhat akin to the penal colonists and indentured servants who populated Australia and some Middle Atlantic American colonies.

Sugarcane was the first crop to be introduced into commercial cultivation in the New World. It is reliably estimated that by 1530, less than forty years after the initial European contact, there were more than a dozen sugar plantations operating in the West Indies.

In his book *Seeds of Change*, Henry Hobhouse writes of the beginning of African enslavement. In 1443 one of Prince Henry's returning captains brought news of a capture at sea of a crew of black Arabs and Moslems:

> These men, who were of mixed Arab-Negro parentage and Moslems, claimed that they were of a proud race and unfit to be bondsmen. They argued forcefully that there were in the hinterland of Africa many heathen blacks, the children of Ham, who made excellent slaves, and who they could enslave in exchange for their freedom. Thus began the modern slave trade—not the transatlantic trade, which was yet to come, but its precursor, the trade between Africa and southern Europe.[4]

Hobhouse goes on to describe sugar slavery in the New World:

> Sugar slavery was of quite a different order. It was the first time since the Roman *latifundia* that mass slavery had been used to grow a crop for trade (not subsistence) in a big way. It was also the first time in history that one race had been uniquely selected for a servile role. Spain and Portugal voluntarily abjured the enslavement of East Indian, Chinese, Japanese or European slaves to work in the Americas.[5]

The slave trade was itself a kind of addiction. The early importation of African slave labor into the New World was for one purpose only, to support an agricultural economy based on sugar. The craze for sugar was so overwhelming that a thousand years of Christian ethical conditioning meant nothing. An outbreak of human cruelty and bestiality of incredible proportions was blandly accepted by the institutions of polite society.

Let us be absolutely clear, sugar is entirely unnecessary to the human diet; before the arrival of industrial cane and beet sugar humanity managed well enough without refined sugar, which is nearly pure sucrose. Sugar contributes nothing that cannot be gotten from some other, easily available source. It is a "kick," nothing more. Yet for this kick the dominator culture of Europe was willing to betray the ideals of the Enlightenment by its collusion with slave traders. In 1800 virtually every ton of sugar imported into England had been produced with slave labor. The ability of the ego-dominator culture to suppress these realities is astonishing.

If it seems that too much ire is vented on the sugar habit, it is because in many ways the addiction to sugar seems a distillation of all the wrongheaded attitudes that attend our thinking about drugs.

SUGAR AND THE DOMINATOR STYLE

When temporal distance from the original partnership paradise increases, when the connection with the vegetable/feminine matrix of planetary life slips far into the past, then the hold of cultural neurosis increases and manifestations of unchecked ego and dominator theories of social organization proliferate. Slavery, almost unknown during the medieval period, when the notion of private property restricted ownership of anything to a privileged few, returned with a vengeance to fill the need for manpower in the labor-intensive colonial cultivation of sugar. Thomas Hobbes's vision of human society as the inevitable subjugation of the weak by the strong and Jeremy Bentham's notion of the ultimate economic basis of all social worth signal that values that seek to nurture the earth and to participate with it in a life of natural emotive balance have been forsaken for the rapacious self-centeredness of Faustian science. The soul of the planet, shrunken by Christian monotheism to the dimensions of a human being, is finally denied any existence at all by the heirs of Cartesian rationalism.

The stage is then set for the evolution of a human self-image that is entirely dis-ensouled, adrift in a dead universe devoid of meaning and without moral compass. Organic nature is seen as war, meaning becomes "contextual," and the cosmos is rendered meaningless. This process of deepening cultural psychosis (an obsession with ego, money, and the sugar/alcohol drug complex) reaches its culmination in the mid-twentieth century with Sartre's appalling assertion that "nature is mute."

Nature is not mute, but modern man is deaf—made deaf because he is unwilling to hear the message of caring, balance, and cooperation that is nature's message. In our state of denial we must proclaim nature mute—how else to avoid facing the awful crimes we have committed for centuries against nature and each other. The Nazis said that Jews were not true human beings and that their mass murder was thus not of any consequence. Some industrialists and politicians use a similar dis-ensouling argument to excuse the destruction of the planet, the maternal matrix necessary to all life.

Only a terminal addiction to the ego and styles of brutal domination could give rise to a mass mental environment in which such statements could appear plausible, let alone true. Sugar stands at a watershed in such matters, for sugar and the caffeine drugs that spread with it reinforce and support industrial civilization's unreflecting emphasis on efficiency at the price of Archaic human values.

THE DRUGS OF GENTILITY

In the opening lines of his magnificent poem "Sunday Morning," Wallace Stevens delivers an image of radiant transcendence and the familiar and ordinary worthy of Cézanne:

> Complacencies of the peignoir, and late
> Coffee and oranges in a sunny chair,
> And the green freedom of a cockatoo
> Upon a rug mingle to dissipate
> The holy hush of ancient sacrifice.[6]

Stevens's lines evoke an aura of genteel satiety that surrounds the drug caffeine. "Sunday Morning" reminds us that our stereotyped notion of what constitutes drugs is strained when we are asked to consider such delicate accessories of bourgeois sensibility as tea, coffee, and cocoa as being in the same category as heroin and cocaine. Yet all are drugs; our unconscious striving to find our way back to the sensory ratios of prehistory has led us to develop countless variations on the act of paying homage to plant-based psychoactivity. Mild stimulants, with nondestructive or manageable impact, have been a part of the diet of primates since long before the emergence of hominids. Caffeine is the alkaloid that lies at the basis of much of the human involvement with plants that stimulate. Caffeine is a powerful stimulator well below the toxic dose. It occurs in tea and coffee and in numerous other plants, such as *Ilex paraguayensis*, the source of maté, or *Paullinia yoco*, an appetite-suppressing Amazonian liana, which have their own localized but ancient and highly ritualized styles of use.

Caffeine is bitter, and the inevitable discovery that it could be made more palatable with the addition of honey or sugar set the stage for the very prevalent and little remarked synergistic effect that occurs between sugar and the various caffeine beverages. Sugar's tendency to become addictive is reinforced if sugar is also being used to make the ingestion of a stimulating alkaloid such as caffeine more palatable.

Sugar is culturally defined by us as a food. This definition denies that sugar can act as a highly addictive drug, yet the evidence is all around us. Many children and compulsive eaters live in a motivational environment primarily ruled by mood swings resulting from cravings for sugar.

COFFEE AND TEA: NEW ALTERNATIVES TO ALCOHOL

For all practical purposes we can say that tea, coffee, and cocoa were introduced simultaneously into England in the 1650s. For the first time in its history, Christian Europe had an alternative to drinking alcohol. All three were stimulants; all were brewed with hot water that had been boiled and thus rendered free from the then-rampant problem of waterborne diseases; and all required copious amounts of sugar. The sugar craze promoted coffee, tea, and chocolate use, which in turn promoted sugar consumption. And the new stimulants were grown in the same colonial holdings then proving so profitable in the production of sugar. Tea, coffee, and cocoa held out the possibility of crop diversification in the colonies and hence of greater economic stability for both colony and mother country.

By 1820 many thousands of tons of tea were being imported into Europe each year, with about 30 million pounds being consumed in the United Kingdom alone. The tea for the European market all came from the southern Chinese coastal city of Canton from the mid-seventeenth to the early nineteenth centuries. The tea buyers were not allowed to penetrate inland, nor were they privy to any of the details of the cultivation and cropping of the tea plant. As Hobhouse writes, "History's joke on Europe is that for nearly two centuries a commodity was imported halfway across the world, and that a huge industry grew up involving as much as 5 percent of England's entire gross domestic product, and yet no one knew anything about how tea was grown, prepared, or blended."[7]

Such ignorance was no barrier to the commercial exploitation of tea; however, the Turkish capture of Constantinople in 1453 certainly was. When trade routes through the eastern Mediterranean were in the hands of the Turk, there was considerable pressure on the sciences of navigation and shipbuilding to perfect the ocean route to the East via the Cape of Africa. The route was discovered, in 1498, by Vasco da Gama.

When Dutch and Portuguese navigators eventually reached the Moluccas, in eastern Indonesia, then called the Spice Islands, spices became much cheaper in Europe and the struggle among all parties to create monopolies was joined. The type of organization best able to maintain a monopoly was the trading company, a group of merchants drawn together to reduce capital risks and competition. The large, well-armed ships of the various East India companies spelled the end of the age of the self-employed merchant-captain. The British East India Company, destined to become the most important of the trading companies, was founded in 1600.

From that date until 1834, when free trade liberals opened the tea trade to all interested parties, the company controlled the tea trade to its great advantage:

> The British East India Company was believed to add at least a third to the price of tea, thus taking £100 a ton out of the 375,000 tons imported during the eighteenth century. This global figure obscures the rise, on the same basis, of the East India Company's cut, from a sum equivalent to $17 million at the beginning of the century to an annual equivalent of $800 million in 1800. The East India Company was big business, hated and loathed by smugglers and consumers alike, and a symbol of corrupt, complacent monopoly.[8]

TEA BREWS A REVOLUTION

Toward the end of the eighteenth century the tea trade was in crisis and the government of Lord North made a series of ill-considered decisions that were not only to ruin the tea trade but were also to lose England her colonies in North America. North's strategy was to sell tea at reduced prices in the colonies, thus diminishing surpluses and driving smuggler competitors out of business. He also sought to place a small and, he imagined, inconsequential tax on tea going to the colonies, simply to force the unruly colonists to submit to imperial authority. As is common knowledge, this tea tax was the straw that broke the camel's back, in the political foment that was then gripping the American colonies. On December 16, 1773, angry colonial radicals in Boston turned on His Majesty's tea ships and destroyed their cargo. The salty tea of revolution was brewed that night. And there were other "tea parties," at New York, Charleston, Savannah, and Philadelphia. The affair might have passed off in a few weeks had not the British response of closing the port of Boston made the Declaration of Independence inevitable.

By the early 1800s the tea trade was showing signs of strain. On the European continent the Napoleonic wars had left coffers depleted. The response had been to print paper money unsecured by gold, and this practice eventually resulted in serious inflation: costs rose, product values rose much less, resulting in economic misery. The panacea to this economic impasse was opium.

EXPLOITATION CYCLES

The opium trade was nothing less than British terrorism waged against the population of China until the Chinese government's restrictions against the importation of opium were totally done away with. There is in these events a pattern that has been repeated in our own century. Just as the dealers of the drug tea turned to opium when their tea market suffered depression, so did Western intelligence groups, such as the CIA and the French secret service, turn their attention to the importation of cocaine in the eighties, after having lost a near monopoly on heroin to the heroin-dealing mullahs of the Iranian Revolution. The history of commercial drug synergies—the way in which one drug has been cynically encouraged and used to support the introduction of others—over the past five hundred years is not pleasant to contemplate. Perhaps that is why the exercise is so rarely undertaken.

The cycles began with sugar. As discussed, sugar, whose existence depended on a savage slave trade, deepened its claim on consumers throughout the sixteenth century. The seventeenth-century introduction of tea, coffee, and chocolate only drove the craze for sugar to new heights. Through its use in caffeine drinks and distilled alcohol, sugar played a major indirect role in furthering the dominator culture's suppression of the underclass and of women of all classes. Slavery to drugs is a tired metaphor, but in the case of sugar the metaphor was made horribly real.

When the tea market collapsed, the distributing system that had been put in place and capitalized by the British East India Company turned to the production and selling of opium and the exploitation of the Chinese population which was outside of the colonial system proper. The invention of morphine (1803) and then heroin (1873) carries us to the threshold of the twentieth century. Alarmed social reformers who attempted to legislate drug use only succeeded in driving it underground. There it remains, controlled today, not by robber baron corporations operating under public charter, but by international crime cartels often posing as intelligence agencies. It is, as William Burroughs has remarked, "Not a pretty picture."

Since the Age of Exploration, drugs and plant products have become increasingly important factors in the equations of international diplomacy. No longer are the distant tropical regions and peoples of the world to languish unattended by the rapacious eye of the white man; they have become production areas populated by an indentured labor force and expected to provide raw materials and a ready market for finished goods. Like the maenads lost in the transport of Dionysian fury, the sugar-intoxicated dominator economies of Europe sought to devour their own children.

COFFEE

The eleventh-century Persian polymath Avicenna, who in 1037 became history's first recorded death from opium overdose, was one of the first to write about coffee, though it had been in use for some time in Ethiopia and Arabia, where the source plant occurred wild. On the Arabian peninsula it has long been known that coffee was a plant of marvelous properties. There is even an apocryphal story that when the Prophet lay ill he was visited by the Archangel Gabriel who offered him coffee to restore him to health. Because of the plant's long association with the Arabs, Linnaeus, the great Danish naturalist and the inventor of modern scientific taxonomy, named the plant *Coffea arabica*.

When coffee was first introduced to Europe, it was used as a food or medicine; the oil-rich berries were pulverized and mixed with fat. Later ground coffee was mixed into wine and cooked to provide what must have been a quite stimulating and intense refreshment. Coffee was not brewed as a drink until around 1100 in Europe, and only in the thirteenth century did the modern practice of roasting coffee beans begin in Syria.

Though coffee was an Old World plant and was used in some circles a long time before tea, nevertheless tea cleared the way for the popularity of coffee. Their stimulant properties made caffeine in coffee and its close cousin theobromine in tea the ideal drugs for the Industrial Revolution: they provided an energy lift, enabling people to keep working at repetitious tasks that demanded concentration. Indeed, the tea and coffee break is the only drug ritual that has never been criticized by those who profit from the modern industrial state. Nevertheless, it is well established that coffee is addictive, causes stomach ulcers, can aggravate heart conditions, and can cause irritability and insomnia and, in excessive doses, even tremors and convulsions.

CONTRA COFFEE

Coffee has not been without its detractors, but they have always been in the minority. Coffee was widely blamed for the death of the French minister Colbert, who died of stomach cancer. Goethe blamed his habitual *caffè latte* for his chronic melancholia and his attacks of anxiety. Coffee has also been blamed for causing what Lewin called "an excessive state of brain-excitation which becomes manifest by a remarkable loquaciousness sometimes accompanied by accelerated association of ideas. It may also be observed in coffee house politicians who drink cup after cup of black coffee and by this abuse are inspired to profound wisdom on all earthly events."[9]

The tendency to excessive raving after coffee drinking apparently lay behind several edicts against coffee issued in Europe in 1511. The prince of Waldeck pioneered an early version of the drug-snitch program when he offered a reward of ten thalers to anyone who would report a coffee drinker to the authorities. Even servants were rewarded if they informed on employers who had sold them coffee. By 1777, however, authorities in continental Europe recognized the suitability of coffee for use by the pillars of dominator society—the clergy and the aristocracy. Punishment for a coffee offense by members of less privileged classes was usually a public caning followed by fines.

And, of course, coffee was once widely suspected of causing impotence:

> It has frequently been stated that the drinking of coffee diminishes sexual excitability and gives rise to sterility. Though this is a mere fable, it was believed in former times. Olearius says in the account of his travels that the Persians

> drink "the hot, black water *Chawae*" whose property it is "to sterilize nature and extinguish carnal desires." A sultan was so greatly attracted by coffee that he became tired of his wife. The latter one day saw a stallion being castrated and declared that it would be better to give the animal coffee, and then it would be in the same state as her husband. The Princess Palatine Elizabeth Charlotte of Orleans, the mother of the dissipated Regent Philip II, wrote to her sister: "Coffee is not so necessary for Protestant ministers as for Catholic priests, who are not allowed to marry and must remain chaste. . . . I am surprised that so many people like coffee, for it has a bitter and a bad taste. I think it tastes exactly like foul breath."[10]

The physician-explorer Rauwolf of Augsburg, who later became the discoverer of the first tranquilizer, the plant extract rauwolfia, found coffee apparently long established and widely traded in Asia Minor and Persia when he visited the area in the mid-1570s. Accounts such as Rauwolf's soon made coffee a fad. Coffee was introduced in Paris in 1643, and within thirty years there were over 250 coffee houses in the city. In the years immediately preceding the French Revolution there were nearly 2,000 coffee establishments operating. If wild talk is the mother of revolution, then certainly coffee and coffee houses must be its midwife.

CHOCOLATE

The introduction of chocolate into Europe is almost no more than a coda to the craze for caffeine stimulation that began with the Industrial Revolution. Chocolate, made from the ground beans of a native Amazonian tree, *Theobroma cacao,* contains only small amounts of caffeine but is rich in caffeine's near-relative theobromine. Both are chemicals with close relatives that occur endogenously in normal human metabolism. Like caffeine, theobromine is a stimulant, and the addictive potential of chocolate is significant.[11]

Cacao trees had been introduced into central Mexico from tropical South America centuries before the arrival of the Spanish *conquistadores*. There they had a major sacramental role in Maya and Aztec religion. The Maya also used cacao beans as the equivalent of money. The Aztec ruler Montezuma was said to be seriously addicted to ground cacao; he drank his chocolate unsweetened in a cold water infusion. A mixture of ground chocolate and psilocybin-containing mushrooms was served to the guests at the coronation feast of Montezuma II in 1502.[12]

Cortes was informed of the existence of cacao by his mistress, a Native American woman named Doña Marina, who had been given to Cortes as one of nineteen young women offered in tribute by Montezuma. Assured by Doña Marina that cacao was a powerful aphrodisiac, Cortes was eager to begin cultivation of the plant; he wrote to the emperor Charles V: "On the lands of one farm two thousand trees have been planted; the fruits are similar to almonds and are sold in a powdered state."[13]

Shortly thereafter, chocolate was imported into Spain, where it was soon extremely popular. Nevertheless the spread of chocolate was slow, perhaps because so many new stimulants were then vying for European attention. Chocolate did not appear in Italy or the Low Countries until 1606; it reached France and England only in the 1650s. Except for a brief period during the reign of Frederick II, when it became the favorite vehicle for poisons used by professional poisoners, chocolate has steadily increased in popularity and annual tonnage produced.

It is extraordinary that in the relatively short span of two centuries four stimulants—sugar, tea, coffee, and chocolate—could have emerged out of local obscurity and become a basis for vast mercantile empires, defended by the greatest military powers ever known to that time and supported by the newly reintroduced practice of slavery. Such is the power of "the cup that cheers, but not inebriates."

12
SMOKE GETS IN YOUR EYES: OPIUM AND TOBACCO

Few plants can lay claim to such complex and tangled relationships to human beings as can the opium poppy and the tobacco plant. Both plants are central to extremely addictive behaviors in human beings that shorten life and burden society with medical and financial consequences. Yet the general attitude toward these plants could hardly be more different. Opium is illegal throughout most of the world. The poppy-growing areas that are the source of raw opium are closely monitored by photo surveillance satellites, and yearly advance projections of world opium production are closely studied by governments to aid them in calculating how much of their budgets to allocate to treatment of addicts, foreign eradication efforts, and domestic interdiction of refined opium products like morphine and heroin.

Tobacco, on the other hand, is probably the most widely consumed plant drug on earth. No nation has decreed smoking of tobacco illegal, and indeed, any country that sought to do so would find itself at loggerheads with one of the most powerful international narcotic cartels ever to exist. Yet there is no dispute that tobacco smoking is the cause of early death for millions of people; lung cancer, emphysema, and heart disease have all been linked to smoking. And tobacco is no less addicting than the supposed hardest of hard drugs, heroin. When this fact was stated by U.S. Surgeon

General C. Everett Koop, it was quickly buried in the storm of derision unleashed by the major American tobacco companies and their legions of addicted customers.

PARADOXICAL ATTITUDES

What can we learn from the comparison of these two plants? Both have a long history of human usage, both are addictive and ultimately destructive, and yet one is firmly integrated into our lifestyles and sold to us as masculine, sophisticated, pleasurable, while the other is illegal, furiously suppressed, inveighed against as suicidal, and viewed with an unreflecting horror that earlier generations reserved for Bolsheviks, suffragettes, and oral sex.

This situation is but another example of the hypocrisy of dominator culture as it picks and chooses the truths and realities that it finds comfortable. The fact is that, while heroin is highly addictive and while one of its preferred routes of ingestion, intravenous self-injection, offers opportunities for the spread of serious disease, it is nevertheless no more dangerous than its legal and highly touted competitor, tobacco: "Volumes of scientific research . . . have concluded that no known organic damage is caused by the use of heroin. It is a physically benign, though powerfully addicting, substance."[1]

The differences in the way that society views these two now globally pandemic plant-based drugs cannot be the result of a reasonable assessment of their deleterious social impacts. If it were, then attitudes toward these two plants would be similar. As it is, we must look at effects unrelated to the shared property of addiction to understand why dominator society has chosen to suppress one and exalt the other.

SMOKING INTRODUCED TO EUROPE

Tobacco is native to the New World, and so is the custom of smoking plant material to obtain narcotic effects from it. Smoking may have been known in the Old World during the Neolithic period; scholarly opinion differs. However, there is no evidence of tobacco smoking being a practice known to any of the historical civilizations of the

Old World until Columbus introduced it following his second voyage to the Americas. Less than a hundred years later, small packets of tobacco were being placed into the graves of Lapland shamans! This gives some notion of how quickly tobacco was able to assert its traditional pattern of usage, even in a society that was completely unfamiliar with it. Tobacco—chewed, snuffed, and smoked—has been with us ever since. By the nineteenth century, tobacco use had been culturally classified in Europe as a "man's prerogative." Successful men were judged by the quantity and quality of cigars they smoked. And tobacco was added to the long list of male dominator privileges which included nearly all alcohol (brandies for the ladies, please), control of finance, access to prostitutes, and control of political power (recall those "smoke-filled rooms").

Even in today's drug-conscious atmosphere, no contradiction is perceived between the strident calls to eliminate the use of drugs in professional athletics and the figure of the tobacco-chewing major league pitcher, eyes hardened with narcotic intensity as he strides to the mound. Does the elimination of drugs from competitive sports mean the extinction of that lovable figure, the lump-cheeked hayseed with a good pitching arm? Somehow I doubt it.

As tobacco was achieving its present stature, opium, too, was enjoying a vogue, albeit nothing on the scale of tobacco. Laudanum, tincture of opium in alcohol, was being used as a cure for colic in infants, a "women's tonic," a cure for dysentery, and most significantly, by writers, travelers, and other Bohemian types, as a stimulant of the creative imagination. Morphine, which must be injected, was the first alkaloid to be isolated. This event, in 1805, cast a dark shadow over the halcyon world of the laudanum enthusiast—for, however much artistic mileage Coleridge and De Quincey obtained from their imagined enslavement to the "fiend of opium," their addictions, though serious, when judged in the light of modern experience with cocaine free-base and new synthetic forms of heroin, appear almost minor.

THE ANCIENT LURE OF OPIUM

The seed of the poppy is a delicious and nonpsychoactive food, as all enthusiasts of poppy seed rolls can attest. Yet when the seed capsule is scratched with blade or fingernail, a milky latexlike material soon accumulates and, as it hardens, turns a dark brown. This material is raw opium. Like the psilocybin mushroom with its association with cattle, and the parasitism of ergot on rye and other cereals, the opium poppy is a major psychoactive plant that has evolved in the presence of a human food source. In the case of the opium poppy, *Papaver somniferum*, the psychoactivity and the nutritional value are sectioned off into different parts of the same plant.

Opium in various forms has been in the physicians' armamentarium since at least 1600 B.C. An Egyptian medical treatise of that period prescribed opium for crying children just as Victorian nannies dosed infants with opiate-laced Godfrey's Cordial to keep them quiet.

For most of its history opium was not smoked but rather the black sticky resin was dissolved into wine and drunk, or else rolled into a pellet and swallowed. Opium, as a cure for pain, as euphorint and rumored aphrodisiac, was known in Eurasia for several thousands of years.

During the waning of the millennia-long Minoan civilization and its religion of Archaic worship of a Great Mother, the original source of the connection to the Goddess of vegetable nature came eventually to be replaced by the intoxication of opium. Early Minoan texts testify to the fact that poppies were widely cultivated on both Crete and Pylos during the Late Minoan; in these texts, the poppy head is used as an ideogram in financial tallies. The yield of poppies indicated is so huge that for some time it was assumed that these numbers must refer to grain rather than opium. The confusion of grain and poppy is easy to understand as Demeter was the goddess of both (see Figure 19). In fact, how much of the lore of the poppy was transferred to the Greek Mysteries of Demeter on the mainland remains to be elucidated, especially as there is some iconographic confusion between the poppy flower and the pomegranate, also a plant associated with the Mysteries. Kerényi quotes Theokritos VII. 157:

> For the Greeks Demeter was still a poppy goddess,
> Bearing sheaves and poppies in both hands.[2]

A remarkable illustration in Erich Neumann's *The Great Mother* shows the Goddess in association with a beehive and holding poppy

FIGURE 19. Demeter with barley, opium, and snakes. Courtesy of Fitz Hugh Ludlow Library.

seed capsules and heads of grain in her left hand while resting her right hand on one of the unadorned pillars central to the Minoan earth religion (see Figure 20). Rarely have so many elements of the Archaic technology of ecstasy been brought together so explicitly. The figure is almost an allegory of the transformation of Minoan shamanic spirituality in its late phase. Its mushroom roots are symbolized in the aniconic column; they are the touchstone of the Goddess who looks toward the promise of poppies and ergotized grain. The hive of bees introduces the theme of honey, the archetypal image of ecstasy, female sexuality, and preservation that survives the shifting botanical identities of the sacraments.

Poppies and gum opium were known to the ancient Egyptians and appear in their funereal arts as well as in the earliest medical papyruses. Poppies were known in several varieties to the Persians; in ancient Greece and elsewhere, the poppy was known as "the destroyer of grief":

> Theophrastus knew of it as a sleep inducing drug in 300 B.C., and his observations were repeated by Pliny in the first

FIGURE 20. Spes, with Sheaves and Hive. From Eric Neumann's *The Great Mother* (New York: Pantheon, 1955), p. 263.

> century A.D. with added thoughts on opium poisoning. The Greeks consecrated the poppy to Nyx, goddess of night, Morpheus, son of Hypnos and god of dreams, and Thanatos, god of death. They summarized all of its properties in the deities to whom it was offered. Opium spread throughout the Islamic world after the seventh century. It was undoubtedly used both as a dysentery cure and for those overburdened with grief and care.[3]

Though opium's habit-forming quality was mentioned by Heraclides of Tarentum, in the third century B.C., this was something even physicians were generally unaware of until nearly two thousand years later. We who have been raised with the notion of addiction as disease may find it hard to believe that chemical dependency upon opiates was not noted or described by medical authorities until early in the seventeenth century. Samuel Purchas, writing in 1613, observed of opium that "but being once used, it must daily be continued on paine of death, though some escape by taking to wine instead." Alethea Hayter comments that "this awareness that opium is addictive is rarely found so early."[4]

For the ancient world, then, opium was that which brought sleep and relief from pain. Opium was prescribed and perhaps overprescribed during the final days of the Roman Empire. Then the use of opium nearly ceased for many centuries in Europe; the early herbals of Saxon England mention juice expelled from poppies as a cure for headache and sleeplessness, but clearly opium played a very minor role in the armamentarium of medieval Europe.[5] Martin Ruland's *Alchemical Lexicon*, published in 1612, mentions only the word "osoror" as a synonym for opium, and then without explanation.

ALCHEMICAL OPIUM

It is to Paracelsus, the famed "father of chemo-therapy," that we can trace the revival of interest in opium. The great sixteenth-century Swiss alchemist, medical reformer, and quack advocated and used opium on a lavish scale. Here again, as in the case of distilled alcohol, it is an alchemist, one involved in the pursuit of the spirit assumed to be locked into matter, who discovered the means to release the power locked within a simple plant. And, like Lully before him, Paracelsus assumed that he had discovered the universal panacea: "I possess a secret remedy which I call laudanum and which is superior to all other heroic remedies."[6]

Shortly after Paracelsus began promulgating the virtues of opium, physicians of his school of thought were preparing nostrums whose sole basis of activity was the copious amount of opium that they contained. One of these enthusiastic followers, the alchemist van Helmont, became well known as "Doctor Opiatus," the first "croaker" or junk doctor.

TOBACCO REDUX

While the "iatro-chemists" of the Paracelsan persuasion were spreading the use of opium in Europe, an exotic newcomer was quietly making its way onto the European stage. Tobacco was the first and most immediate payoff of the discovery of the New World. On November 2, 1492, less than a month after his first arrival in the New World, Columbus landed on the north coast of Cuba. There the Admiral of the Ocean Sea dispatched two gift-laden members of his crew to the interior of the island, where he believed that the king of the many coastal villages he had seen must reside. Doubtless there was still some hope in the admiral's mind that his men would return with word of gold, precious stones, fine woods, and spices—the wealth of the Indies. Instead, the scouts returned with an account of men and women who partially inserted burning rolls of leaves into their nostrils. These burning rolls were called *tobacos* and consisted of dry herbs wrapped up in a large dry leaf. They were lit at one end, and the people sucked at the other and "drank the smoke," or inhaled, something that was utterly unknown in Europe.

De las Casas, the bishop of Chiapas, who published the account of Columbus's in which this description occurs, added this observation:

> I know of Spaniards who imitate this custom, and when I reprimanded the savage practice, they answered that it was

> not in their power to refrain from indulging in the habit. Although the Spaniards were extremely surprised by this peculiar custom, on experimenting with it themselves they soon obtained such pleasure that they began to imitate the savage example.[7]

Four years after the first voyage, the hermit Romano Pane, whom Columbus had left in Haiti at the conclusion of the second voyage to the New World, described in his journal the native habit of inhaling the tobacco fumes with the help of a bird-bone instrument inserted into the nose and held over tobacco strewn on a bed of coals. The consequences of this simple ethnographic observation have still to be calculated. It introduced into Europe an extremely efficient method for conveying drugs—including many potentially dangerous drugs—into the human body. It made possible the worldwide pandemic of tobacco smoking. It was a fast-acting and easily abused route of administering both opium and hashish. And it was the distant ancestor of the crack cocaine and PCP smoker. It also, it must be said, makes possible the most profound of the indole-hallucinogen-induced ecstasies, the rarely encountered but incomparable practice of smoking dimethyltryptamine.

SHAMANIC TOBACCOS

Tobacco smoking was widespread in North America at the time of the European contact. While the habit of taking hallucinogenic DMT-containing snuffs was also prevalent in the Caribbean cultural area, there are no confirmed reports of materials other than tobacco being smoked.

The high culture of the Maya that flourished until the mid-800s in Mesoamerica had an old and complex relationship with tobacco and the habit of smoking it. The tobacco of the Classical Maya was *Nicotiana rustica*, which is still in use among aboriginal populations in South America today. This species is much more potent, chemically complex, and potentially hallucinogenic than the commercial grades of *Nicotiana tabacum* available today. The difference between this tobacco and cigarette tobacco is profound. This wild tobacco was cured and rolled into cigars which were smoked. The trancelike state that followed, partially synergized by the presence of compounds that included MAO inhibitors, was central to the shamanism of the Maya. Recently introduced antidepressants of the MAO inhibitor type are distant synthetic relatives of these natural compounds. Francis Robicsek has published extensively on the Mayan fascination with tobacco and its chemical complexity:

> It also must be recognized that nicotine is by no means the only bioactive substance in the tobacco leaf. Recently alkaloids of the harmala group, harman and norharman, have been isolated from cured commercial tobaccos and their smoke. They constitute a chemical group of beta-carbolines, which include harmine, harmaline, tetrahydroharmine, and 6-methoxy harmine, all with hallucinogenic properties. While to date no native varieties of tobacco have been analyzed for these substances, it is a reasonable supposition that their composition may vary widely, depending upon the variety and growth, and that some of the native-grown tobaccos may contain a relatively high concentration of them.[8]

Tobacco was and is the ever-present adjunct of the more powerful and visionary hallucinogenic plants wherever in the Americas they are used in a traditional and shamanic way.

And one of the traditional uses of tobacco involved the New World's invention of the first enemas. Peter Furst has researched the role of enemas and clysters in Mesoamerican medicine and shamanism:

> It has only recently come to light that the ancient Maya like the ancient Peruvians employed enemas. Enema syringes or narcotic clysters, and even enema rituals, were discovered to be represented in Maya art, an outstanding example being a large painted vase dating A.D. 600–800, on which a man is depicted carrying an enema syringe, applying an enema to himself, and having a woman apply it to him. As a result of this newly discovered scene, archaeologist M. D. Coe was able to identify a curious object held by a jaguar deity on another painted Maya vessel as

an enema syringe. If the enemas of the ancient Maya were, like those of the Peruvian Indians, intoxicating or hallucinogenic, they might have consisted of fermented *balché* (honey mead). *Balché* is a very sacred beverage and it may have been fortified with tobacco or with morning-glory seed infusions. Datura infusions and even hallucinogenic mushrooms may have been taken in this way. Of course they could also have used a tobacco infusion alone.[9]

TOBACCO AS QUACK MEDICINE

Any drug introduced into use inevitably winds up associated with a number of quack medical theories and treatments. Cocaine abuse, as we shall see, was preceded by the craze for the tonic Vin de Mariani, and heroin was touted as a cure for morphine addiction. Lest we recoil from the enema rituals of the Maya, consider that in 1661 the Danish physician Thomas Bartholin recommended not only tobacco-juice enemas but also tobacco-smoke enemas to his patients:

> Those who have swallowed tobacco by accident can testify to its purgative effect. This property is employed in the tobacco clyster used as an enema. My beloved brother, Erasmus, has shown me the method. Smoke from two pipes [filled with tobacco] is blown into the intestines. A suitable instrument for this was devised by the ingenious English.[10]

Not to be outdone by the clever English, an eighteenth-century French physician named Buc'hoz advocated the use of "intravaginal insufflation of tobacco smoke to cure hysteria."

Quite aside from these eccentric and bizarre applications of the use of tobacco, and despite the scorn of the clergy, the habit of smoking spread quickly in Europe. Every drug, during the process of its introduction into a new cultural milieu, is hailed as a "love drug," this apparently being the most effective of all advertising ploys. Drugs as diverse as heroin and cocaine, LSD, and MDMA were all at some point presented as promoting intimacy, sexual or psychological. Tobacco was no different; part of the reason for its rapid spread was widely circulated sailors' yarns of its remarkable properties as an aphrodisiac:

> Sailors told of the women of Nicaragua who smoked this weed and displayed an ardor undreamed of. It was probably this rumor that clinched the popularity of smoking among the women of Europe. Perhaps this is the reason why an ex-Franciscan monk, André Thevet, experienced such success in introducing tobacco to the French court in 1579.[11]

Thevet fully intended that tobacco be smoked and used as a recreational drug. Earlier the French ambassador to Portugal, Jean Nicot, had experimented with crushed tobacco leaves used as a snuff for the purpose of curing migraine headache. In 1560 Nicot conveyed a sample of his snuff to Catherine de Medici, who was a chronic migraine sufferer. The queen was enthusiastic about the plant's powers, and it briefly became known as "Herba Medicea" or "Herba Catherinea." Nicot's snuff was derived from the generally more toxic *Nicotiana rustica*, the classic, shamanic tobacco of the Maya. Thevet's *Nicotiana tabacum* conquered Europe in the form of the cigarette and was the plant that became the basis for the tremendously important tobacco economy that grew up in the colonial New World.

CONTRA TOBACCO

Tobacco was not welcomed by all. Pope Urban VIII ordered excommunication for anyone who smoked or used snuff in the churches of Spain. In 1650 Innocent X forbade snuff taking in the basilica of St. Peter, on pain of excommunication. Protestants also decried the new habit, and were led in their effort by no less than King James I of England, whose inflammatory *Counterblaste to Tobacco* appeared in 1604:

> And now good Country man, let us (I pray you) consider what honor or policie can move us to imitate the slavish Indians, especially in so vile and stinking custome. . . . I say without blushing, (why do we) abase ourselves so farre, as

> to imitate these beastly Indians, slaves to the Spaniards, refuse to the world, and as yet aliens to the holy Covenant of God? Why do we not as well imitate them in walking naked as they doe? . . . Yes, why do we not denie God and adore the Devill, as they doe.[12]

Having unleashed this rhetorical "counterblaste," in what may be seen as the first unlimbering of the "just say no" approach, the king turned his attention to other matters. Eight years later a report claimed that in the city of London alone, there were no fewer than 7,000 tobacconists and tobacco houses! Tobacco smoking and snuff taking were pursued at the level of intensity of a modern craze.

TOBACCO TRIUMPHANT

In commercial terms, tobacco did not attain major importance until after the close of the Thirty Years War in 1648. By then American colonies were in place and well able to participate in the mercantile economy that had been established. In fact, this economy ran in large part on the tobacco of the North American colonies and the distilled alcohol and raw sugar of the more tropical outposts. The Age of Enlightenment was firmly founded on a drug-based economy.

A remarkable process attended the introduction of tobacco into Europe: because of the emphasis on the recreational potential and the large-scale planting of *Nicotiana tabacum*, the less toxic of the two major species, tobacco lost its connotation as a plant of shamanic and even hallucinogenic power. This was more than a matter of shifts in the standard dose and the method of administration. The native tobaccos that I have experienced among Amazonian peoples were extremely disorienting and barely subtoxic. They were definitely capable of producing an altered state of consciousness. The tobacco-using habit as it evolved in Europe was secular and recreational, and hence much more mild strains of tobacco were commercially successful.

Once a drug is discovered, it often goes through a process of dilution before a general consensus is reached on the most desirable level of effect. Moving from eating opium or hashish to smoking of these substances was such a process, as was the move from large doses of LSD in the 1960s to the current practice of taking small doses of LSD for recreational reasons. This latter move may have been the consequence of the small but persistent percentage of people who suffered severe psychotic breaks after using large doses of LSD. The notion of the "correct" dose of a drug is something that a culture evolves over time. (There are, of course, some counterexamples as well; the trend from snuffing powdered cocaine to smoking crack cocaine exemplifies a movement toward larger doses and more dangerous use patterns.)

THE OPIUM WARS

It was the prohibition of tobacco smoking in China by the last emperor of the Ming Dynasty (1628–1644) that led frustrated tobacco addicts to experiment with smoking opium. Before that time the smoking of opium was not known. Thus it is that the suppression of one drug seems inevitably to lead to involvement with another. By 1793 opium and tobacco were being routinely smoked together throughout China.

Beginning in 1729 the Chinese had strictly prohibited the importation and sale of opium. In spite of this, opium imports, brought by the Portuguese from plantations in Goa, continued to rise, until by 1830 more than 25,000 chests of opium were being illegally imported into China. English financial interests that felt threatened by the prohibitions manipulated the situation into the so-called Opium Wars of 1838–1842:

> The East India Company and the British government rationalized the opium trade with the kind of bland hypocrisy which has made the English establishment a byword for three centuries. There was no direct connection between the opium trade and the East India Company which, of course, had a monopoly position in the British tea trade until 1834. . . . The opium was sold at auction in Calcutta. After this the Company abjured all responsibility for the drug.[13]

The incident that triggered this episode of capitalistic terrorism and true drug enslavement on a mass scale was the destruction of 20,000 chests of opium by Chinese authorities. In 1838 the emperor Tao-Kwang sent an official emissary named Lin to Canton to end the illegal trade in opium. Official orders were issued to the British and Chinese drug dealers to remove their goods, but the orders were ignored. Commissioner Lin then burned the Chinese warehouses on land and the British ships awaiting unloading in the harbor. More than a year's supply of opium was sent up in smoke; chroniclers who witnessed the event recalled that the aroma was incomparable.[14]

The controversy dragged on, but eventually, in 1840, war was declared. The British took the initiative, secure in the power and preeminence of the Royal Navy. The Chinese did not have a chance; the war was short and decisive. In 1840 Chusan was captured, and the following year the British bombarded and destroyed forts on the Canton River. The local Chinese commander, Ki Shen, who had succeeded Commissioner Lin, agreed to cede Hong Kong and pay an indemnity of 6 million Chinese silver dollars, worth about £300,000. When the news reached Peking, the emperor was left with no course but to agree. Thus the Chinese suffered considerable losses in money and territory.[15]

Fifteen years later a second war broke out. This war, too, ended unsuccessfully for China. Shortly afterward the Treaty of Tientsin legalized the Chinese opium traffic.

In many ways this incident was to be the model for much larger forays into international drug trading on the part of twentieth-century governments. It showed clearly that the potential marketability of new drugs can and will overwhelm institutional forces that oppose or appear to oppose the new commodity. The pattern established by England's nineteenth-century opium diplomacy has been repeated, albeit with some new wrinkles, in Central Intelligence Agency collusion in the international heroin and cocaine trade of our own time.

OPIUM AND CULTURAL STYLE: DE QUINCEY

In the early nineteenth century, opium was influencing more than the Far Eastern policy of the mercantile empires; it was also having an unexpected influence on the esthetic forms and styles of European thought. In a sense European society was awakening from narcissistic preoccupation with Renaissance Classicism and finding itself a spectator at the seductively metaphysical and esthetically exotic banquet being conducted by the Grand Turk of the Ottomans—a banquet whose major aperitif was the opium vision.

There is no way to avoid a discussion of Thomas De Quincey at this point. Like Timothy Leary in the 1960s, De Quincey was able to convey the visionary power of what he experienced. For De Quincey this was a power imprisoned within the labyrinth of the poppy. He was able to convey the opium vision with the force and the filigree of melancholia typical of Romanticism. Almost singlehandedly he created, in his *Confessions of an English Opium-Eater*, the cultural image, the Zeitgeist, of the experience of opium intoxication and a metaphysic of opium. He invented the form of the drug "confession," the primary genre of subsequent drug literature. His descriptions of the world view of the opium user are unsurpassed:

> Many years ago, when I was looking over Piranesi's "Antiquities of Rome," Mr. Coleridge, who was standing by, described to me a set of plates by that artist, called his "Dreams," and which record the scenery of his own visions during the delirium of a fever. Some of them (I describe only from memory of Mr. Coleridge's account) represented vast Gothic halls, on the floor of which stood all sorts of engines and machinery, wheels, cables, pulleys, levers, catapults, etc., expressive of enormous power put forth, and resistance overcome. Creeping along the sides of the wall, you perceived a staircase; and upon it, groping his way upward, was Piranesi himself. Follow the stairs a little farther, and you perceive it to come to a sudden, abrupt termination, without any balustrade, and allowing no step onward to him who had reached the extremity, except into the depths below. Whatever is to become of poor Piranesi,

> you suppose, at least that his labors must in some way terminate here. But raise your eyes, and behold a second flight of stairs still higher; on which again Piranesi is perceived, by this time standing on the very brink of the abyss. Again elevate your eyes, and a still more aërial flight of stairs is beheld; and again is poor Piranesi busy on his aspiring labors; and so on, until the unfinished stairs and Piranesi both are lost in the upper gloom of the hall. With the same power of endless growth and self-reproduction did my architecture proceed in dreams.[16]

Opium exhilarates the spirit; it can produce endlessly unraveling streamers of thought and rhapsodic speculation. The fifty years following De Quincey's *Confessions* were to see a deep grappling with the impact of opium use on creativity, especially literary creativity. De Quincey pioneered this effort; he was the first writer

> to study deliberately, from within his personal experience, the way in which dreams and visions are formed, how opium helps to form them and intensifies them and how they are then re-composed and used in conscious art—in his case in "impassioned prose," but the process would also apply to poetry. He learned his waking technique as a writer partly from observation of how the mind works in dreams and rêveries under the influence of opium.
>
> It was his belief that opium dreams and rêveries could be in themselves a creative process both analogous to, and leading to, literary creation. He used dreams in his writing not as decoration, not as allegory, not as a device to create atmosphere or to forestall and help on the plot, not even as intimations of a higher reality (although he believed that they were that) but as a form of art in themselves. His study of the workings of the imagination in sleep to produce dreams was pursued with as much concentration as some of his contemporaries devoted to the working of the waking imagination to produce poetry.[17]

FIGURE 21. *La Morphiniste* by Eugene Grassett, 1893. Courtesy of Fitz Hugh Ludlow Library.

THE BEGINNING OF PSYCHOPHARMACOLOGY

The analytical and psychological interests of men like De Quincey and the French psychiatrist J. J. Moreau de Tours, and their attitudes toward the substances that they sought to explore, signify the beginnings of the less than happy effort of science to come to terms with these materials. Implicit in their work is the assumption that intoxication can mimic madness, a strong hint that madness, that most "mental" of maladies, was rooted in physical causes. The opium dream was seen as a kind of waking theater of the imagination. And there is in the fascination with dreams an anticipation of the psychoanalytical methods of Freud and Jung; this fascination is felt throughout the literature of the nineteenth century—in Goethe, in Baudelaire, in Mallarmé, Huysmans, and Heine. It is the sirens' song of the unconscious, silent since the destruction of Eleusis but expressed in Romanticism and the pre-Raphaelites as a pagan exuberance, driven as often as not by devotion to opium. The heavy-lidded harlots of a Beardsley seraglio or the darker labyrinthine visions of Odilon Redon or Dante Gabriel Rossetti epitomize this esthetic.

As the esthetic had a darker side, so too the chemistry of the poppy began to yield more consuming and more virulently addictive derivatives. The hypodermic syringe was discovered in 1853, and from then on the users of opiates have had the cautionary example of the severely addicted intravenous user of morphine to temper their devotion. (See Figure 21.)

The nineteenth century experienced a sorting out of the bewildering variety of new drugs and stimulants that the previous two centuries of exploration and exploitation of far-flung lands had brought. Tobacco use in its various forms became widespread in all social classes, especially among men. Opium was abused by smaller but nonetheless vast numbers of persons, also drawn from all classes. Distilled alcohol was produced and abused in far larger quantities than ever before. In this environment temperance organizations also emerged, and the modern positions vis-à-vis the drug question began to develop. Yet the true impact of the spreading habits of synthetic drug abuse still lay ahead, in the twentieth century.

13

SYNTHETICS: HEROIN, COCAINE, AND TELEVISION

Morphine was isolated in 1805 by the young German chemist Friedrich Sertürner. For Sertürner, morphine was the purest essence of the poppy plant; he named it after Morpheus, the Greek god of dreams. It was this success in isolating the essence of the opium poppy that inspired chemists to attempt the isolation of pure compounds from other proven materia medica. Drugs for the relief of heart disease were isolated from foxglove. Quinine was extracted from the cinchona tree, purified, and used in the colonial conquest of the malarial zone. And from the leaves of a South American bush was extracted a new and promising local anesthetic—cocaine.

Morphine use was restricted and sporadic until after the middle of the nineteenth century. At first its major nonmedical use was as a vehicle of suicide, but this phase was brief and soon morphine was established as a new and very different sort of drug. In 1853 Alexander Wood invented the hypodermic syringe. Before its invention, physicians had used the hollow stems of the lilac plant to introduce drugs inside the body. The syringe arrived just in time to be used to inject morphine into soldiers wounded in the American Civil War and the Franco-Prussian War. This established a pattern that we will meet again in the history of opiates—the pattern of war as vector of addiction.

By 1890 use of morphine on the battlefield had resulted in significant addict populations in both Europe and the United States. So many Civil War veterans returned home as addicts to injectable morphine that yellow journalists referred to morphine addiction as "the soldier's disease."

HARD NARCOTICS

Distilled alcohol and white sugar had preceded morphine as examples of high purity addictive compounds, but morphine set the pattern for the modern "hard drugs," meaning highly addictive injectable narcotics. At first such drugs were derived from opiates, but all too soon cocaine joined the list. Once heroin, invented as a cure for morphine addiction, was introduced, it quickly replaced morphine as the synthetic opiate of choice among addicts. Heroin has retained this position throughout the twentieth century.

Heroin also quickly replaced all other drugs in the public fantasy concerning the evils of drug addiction. Even to this day, with statistics showing that alcohol kills ten times more often than heroin, heroin addiction is still viewed as the depths of drug depravity. There are two reasons for this view.

One reason is the actual addictive power of heroin. The craving for heroin and the illegal or violent acts that the craving may induce have given heroin the reputation as a drug whose addicts will kill for it. Tobacco addicts might kill for their fix, too, if they had to, but instead they simply walk out to a 7-Eleven to buy cigarettes.

The other reason for the distaste with which heroin addiction is viewed is the characteristics of the intoxicated state. Immediately after his shot the heroin addict is cheerful, almost ebullient. This active response to the shot quickly gives way to the "nod" or "nodding out." The junkie's goal with each shot of junk is to "get the nod on," to get into the detached state of twilight sleep in which the long reveries of the opiates can unfurl themselves. In this state there is no pain, no regret, no distraction, and no fear. Heroin is the perfect drug for anyone who has been damaged by lack of self-esteem or traumatized by historical upheaval. It is a drug of battlefields, concentration camps, cancer wards, prisons, and ghettos. It is the drug of the resigned and the dissolute, the surely dying and the victims unwilling or unable to fight back:

Junk is the ideal product . . . the ultimate merchandise. No sales talk necessary. The client will crawl through a sewer and beg to buy. . . . The junk merchant does not sell his product to the consumer, he sells the consumer to his product. He does not improve and simplify his merchandise. He degrades and simplifies the client. He pays his staff in junk.

Junk yields a basic formula of "evil" virus: *The Algebra of Need*. The face of "evil" is always the face of total need. A dope fiend is a man in total need of dope. Beyond a certain frequency need knows absolutely no limit or control. In the words of total need: "*Wouldn't you?*" Yes, you would. You would lie, cheat, inform on your friends, steal, do *anything* to satisfy total need. Because you would be in a state of total sickness, total possession, and not in a position to act in any other way. Dope fiends are sick people who cannot act other than they do. A rabid dog cannot choose but bite.[1]

COCAINE: THE HORROR OF THE WHITENESS

Like heroin, cocaine is a modern high-purity drug derived from a plant with a long history of folk use. For millennia the peoples of the montane rain forests of South America have held cultural values that promote the ritual and religious use of the stimulant/food coca.

Locals in areas where coca has traditionally been cultivated and used will immediately tell one, "*Coca no es un droga, es comida.*" Coca is not a drug, it is food. And indeed this appears to be largely the case. The self-administered doses of ground coca dust contain a significant percentage of the daily requirement of vitamins and minerals.[2] Coca also is a powerful appetite suppressant. The importance of these facts cannot be appreciated without an understanding of the situation regarding protein availability in the Amazonian forest and Andean Altipano. The casual traveler might suppose that the lushness of the tropical forest signifies an abundance of fruits, edible seeds, and roots. This is not the case. Competition for available protein resources is so fierce among the thousands of species of life that comprise the jungle biota that nearly all usable organic materials are actually bound in living systems. Human penetration into such an environment is greatly aided by an appetite-suppressing plant.

Of course appetite suppression is only one characteristic of coca use. The important characteristic is stimulation. The climaxed rain forest is a difficult place to inhabit. Gathering food and building shelter often requires carrying large amounts of material over considerable distances. Often the machete is the only tool to hold the rain forest at bay.

To the ancient Inca culture of Peru, and later to the indigenous people and the mestizo *colonistas*, coca was a goddess, a kind of New World echo of Graves's white goddess Leucothea. Significantly, the goddess Mama Coca as a young girl, offering the saving branch of coca to the Spanish conqueror, figures prominently in the frontispiece of W. Golden Mortimer's classic *History of Coca: The Divine Plant of the Incas* (see Figure 22).

In 1859 cocaine was isolated for the first time. Pharmacology was undergoing a kind of renaissance, and research with cocaine was vigorously pursued over the next several decades. At this point in our discussion it seems hardly necessary to mention that cocaine was first hailed as an obvious cure for morphinism! Medical researchers who were attracted to the new drug included the young Sigmund Freud:

> At present it is impossible to assess with any certainty to what extent coca can be expected to increase human mental powers. I have the impression that protracted use of coca can lead to a lasting improvement if the inhibitions manifested before it is taken are due only to physical causes or to exhaustion. To be sure, the instantaneous effect of a dose of coca cannot be compared with that of a morphine injection; but, on the good side of the ledger, there is no danger of general damage to the body as is the case with the chronic use of morphine.[3]

Freud's findings, which he would later repudiate, were neither very widely publicized nor well received where they were noticed. It was a fellow student of Freud's in Vienna, Carl Koller, who took

FIGURE 22. Mama Coca as a New World goddess who welcomes the arriving Spaniards. From the frontispiece of W. G. Mortimer's *History of Coca: The Divine Plant of the Incas* (San Francisco: And/Or Press, 1974). Courtesy of Fitz Hugh Ludlow Library.

the next step in the medical application of cocaine, the discovery of its use as a local anesthetic. Overnight Koller's discovery revolutionized surgery; by 1885 cocaine was being hailed as a tremendous medical breakthrough. However, as its use spread, its action as an addiction-inducing stimulant was also noted. Cocaine was the inspiration for the unnamed drug that causes sudden personality change in Robert Louis Stevenson's *The Strange Case of Dr. Jekyll and Mr. Hyde*—a fact that contributed to its fast-accruing reputation as a virulent new vice of the wealthy and depraved.

PRO COCAINE

Not all literary references to cocaine portrayed it in such a horrific light. In 1888 British physician Sir Arthur Conan Doyle wrote a now-famous short novel, *The Sign of Four,* in which his detective, the redoubtable Sherlock Holmes, comments on his use of cocaine: "I suppose that its influence is physically a bad one. I find it, however, so transcendingly stimulating and clarifying to the mind that its secondary action is a matter of small amount."[4]

Coca followed the pattern already set with coffee, tea, and chocolate; that is, it quickly attracted entrepreneurial attention. Chief among those who saw commercial opportunities in coca was a Frenchman, M. Angelo Mariani. In 1888 the first bottle of Vin Mariani was marketed (see Figure 23), and soon there was an entire line of coca-based and -laced wines, tonics, and elixirs:

> Mariani was the greatest exponent of the virtues of coca the world has ever known. He steeped himself in coca lore, surrounded himself with Incan artifacts, cultivated a coca garden at his home, and directed a merchandising empire that featured his tonic wine. Through his genius for advertising he came closer to "turning on the World" than any man who ever lived. Queen Victoria, Pope Leo XIII, Sarah Bernhardt, Thomas Edison, and hundreds of other celebrities and medical men gave public testimony to the tonic properties of his products in a series of twelve volumes published by his company.[5]

FIGURE 23. Advertisement for Vin Mariani. Courtesy of Fitz Hugh Ludlow Library.

MODERN ANTIDRUG HYSTERIA

In the United States at the turn of the century, racist rumormongering fanned the hysterical fear that southern blacks, maddened by cocaine, might attack whites. In 1906 the Pure Food and Drug Act was passed; it made cocaine and heroin illegal and set the stage for the legally sanctioned suppression of the synthetic and addictive compounds found in the opium poppy and the coca bush. In contrast to tobacco, tea, and coffee, which were initially resisted and then made legal, morphine/heroin and cocaine began their career in modern society as legal substances but once recognized as addictive were suppressed. Why these drugs and not others? Was the addiction more virulent? Was the use of the hypodermic injection somehow offensive? Or was there some difference in the social and psychological effect of these drugs that made them scapegoats for the damage being done to society by alcohol and tobacco? These are difficult questions, not amenable to easy answers. Yet, if we are to understand the different climate of drug markets and drug use in the twentieth century, these are the questions we must attempt to answer.

Part of the answer may lie in the fact that by the beginning of the twentieth century nearly a hundred years of experience with the social consequences of addictive synthetic drugs was behind us. The cheerful folly of hailing each new pharmacological discovery as a universal panacea had been amply demonstrated. What could be ignored or left undocumented in the eighteenth or even nineteenth century could not be so easily hidden in the twentieth. Ever more rapid communication and transportation networks spread information about the drugs as well as the drugs themselves (Figure 24).

These technologies helped lead to efficiently organized and administered large-scale criminal syndicates. Yet the rise of these syndicates and of modern narcotics production and distribution systems also required connivance on the part of governments. Hard drug addiction had given the drug trade a blackened reputation. Governments that had dealt drugs with impunity for centuries suddenly found themselves, in the new atmosphere of temperance and social reform, forced to legislate this lucrative trade out of the realm of ordinary commerce and into the status of an illicit activity. Governments would now make their drug money in kickback schemes and in situations in which they would be paid to "look the other way."

DRUGS AND GOVERNMENTS

Government involvement in and direct responsibility for the drug trade would diminish, with protection rackets replacing direct earn-

FIGURE 24. *Cocaine Lil* by John Powys. Courtesy of Fitz Hugh Ludlow Library.

ings, while retail prices would rise astronomically. The new price structure made the drug money pie large enough for all parties to profit handsomely—governments and criminal syndicates alike.

In effect, the modern solution has been for the drug cartels to operate as proxies for national governments in the matter of supplying addictive narcotics. Governments can no longer participate openly in the world narcotic trade and claim legitimacy. Only pariah governments operate without fronts. Legitimate governments prefer to have their intelligence agencies cut secret deals with the drug mafiosi while the visible machinery of diplomacy seems all aflutter over the "drug problem"—a problem always presented in such terms as to convince any reasonable person of its utter insolubility. It is significant that the major production areas of hard narcotics are "tribal zones." Modern imperialists would have us believe that, try as they might, they have never been able to overrun and control these areas, in Pakistan and Burma for example, where major production of opium occurs. Consequently, faceless tribal leaders, ever changing and with unpronounceable names, can be held responsible for it all.

From 1914 until World War II, drug distribution was largely in the same hands of the gangsters who directed other illicit operations that characterize gangster subculture: prostitution, loan sharking, and various rackets. The prohibition of alcohol in the United States had created a vast windfall market for hard narcotics, as well as offering the opportunity for easy profits from alcohol manufactured illegally and sold untaxed.

Government manipulation of drug markets occurred elsewhere, too. During World War II the Japanese occupiers of Manchuria took a page from the book of British colonial oppression of a century earlier and produced vast amounts of opium and heroin for distribution inside China. This was done, not with an eye to profit, as in the British case, but with the intent of creating so many addicts that the will of the Chinese people to resist the occupation effectively would be broken. Later, during the 1960s, the Central Intelligence Agency would use the same technique to smother political dissent in American black ghettos under an avalanche of No. 4 China White—heroin of extraordinary purity.[6]

DRUGS AND INTERNATIONAL INTELLIGENCE

The virulence of addictions to synthetics like heroin and cocaine could not long escape the attention of the inheritors of the slave trade and the opium wars—international intelligence agencies and secret police organizations. These shadowy groups have an insatiable need for untraceable money to fund the private armies, terrorist cells, coups d'état, and front groups that are their stock in trade.

Involvement in, and indeed domination of, the world narcotics trade has proven irresistible to groups such as the CIA, Opus Dei, and the French secret service:

> The U.S. Government's Mafia and narcotics connection goes back, as is well known, to World War II. Two controversial joint operations between OSS (Office of Strategic Services) and ONI (U.S. Naval Intelligence) established contacts (via Lucky Luciano) with the Sicilian Mafia and (via Tai Li) with the dope-dealing Green Gang of Tu Yueh-Sheng in Shanghai. Both connections were extended into the post-war period.[7]

The involvement of legitimate institutions remains the same with certain exceptions. In the late 1970s, there was a move in American hard drug culture from emphasis on heroin to emphasis on cocaine. This move was in part a logical consequence of the American military defeat in Vietnam and retraction from Southeast Asia. It was soon reinforced when the Reagan agenda of contra support and narcoterrorism opened new frontiers for covert operations.

Yet it is unlikely that the virulence or social cost of the cocaine epidemic was ever anticipated. Perhaps no one ever asked the question "What are the consequences of hooking the American public on cocaine?" Perhaps the development of smokable, more efficient, and more addictive crack cocaine was unexpected. It is highly likely that the phenomenon of crack is an instance of technology having escaped from the control of its creators. In the 1980s cocaine assumed a form more virulent than any of its earlier victims and detractors could have possibly imagined.

This is a new and disturbing pattern in the evolution of human-drug interactions—a pattern that cannot be ignored. If today we are confronted by a superaddictive form of cocaine, why not tomorrow a superaddictive form of heroin? In fact, such forms of heroin already exist. Fortunately they are simply not as easy to manufacture as is crack cocaine. Ice, a smokable form of highly addictive methamphetamine, has appeared in the drug underground. There will be other drugs in the future—more addictive, more destructive than anything now possible. How, then, will law and society respond to this phenomenon? It is to be hoped the response won't be one of self-righteously holding the addicts up as examples of contemptible behavior.

From a historical point of view, restricting the availability of addictive substances must be seen as a peculiarly perverse example of Calvinist dominator thought—a system in which the sinner is to be punished in this world by being transformed into an exploitable, hapless customer, who is punished for addiction by being relieved of his cash, by the criminal/governmental combine that provides the addictive substances. The image is more horrifying than that of the serpent that devours itself—it is once again the Dionysian image of the mother who devours her children, the image of a house divided against itself.

ELECTRONIC DRUGS

In his science fiction novel *The Man in the High Castle,* Philip K. Dick imagined an alternative world in which World War II had been won by the Japanese and the Third Reich.[8] In Dick's fictional world, the Japanese occupation authorities introduced and legalized marijuana as one of their first moves at pacifying the population of California. Things are hardly less strange here in what conventional wisdom lightheartedly refers to as "reality." In "this world," too, the victors introduced an all-pervasive, ultra-powerful society-shaping drug. This drug was the first of a growing group of high-technology drugs that deliver the user into an alternative reality by acting directly on the user's sensorium, without chemicals being introduced into the nervous system. It was television. No epidemic or addictive craze or religious hysteria has ever moved faster or made as many converts in so short a time.

The nearest analogy to the addictive power of television and the transformation of values that is wrought in the life of the heavy user is probably heroin. Heroin flattens the image; with heroin, things are neither hot nor cold; the junkie looks out at the world certain that whatever it is, it does not matter. The illusion of knowing and of control that heroin engenders is analogous to the unconscious assumption of the television consumer that what is seen is "real" somewhere in the world. In fact, what is seen are the cosmetically enhanced surfaces of products. Television, while chemically noninvasive, nevertheless is every bit as addicting and physiologically damaging as any other drug:

> Not unlike drugs or alcohol, the television experience allows the participant to blot out the real world and enter into a pleasurable and passive mental state. The worries and anxieties of reality are as effectively deferred by becoming absorbed in a television program as by going on a "trip" induced by drugs or alcohol. And just as alcoholics are only vaguely aware of their addiction, feeling that they control their drinking more than they really do . . . people similarly overestimate their control over television watching. . . . Finally it is the adverse effect of television viewing on the lives of so many people that defines it as a serious addiction. The television habit distorts the sense of time. It renders other experiences vague and curiously unreal while taking on a greater reality for itself. It weakens relationships by reducing and sometimes eliminating normal opportunities for talking, for communicating.[9]

THE HIDDEN PERSUADER

Most unsettling of all is this: the content of television is not a vision but a manufactured data stream that can be sanitized to "protect" or impose cultural values. Thus we are confronted with an addictive and all-pervasive drug that delivers an experience whose message is whatever those who deal the drug wish it to be. Could anything provide a more fertile ground for fostering fascism and totalitarianism than this? In the United States, there are many more televisions than households, the average television set is on six hours a day, and the average person watches more than five hours a day—nearly one-third their waking time. Aware as we all are of these simple facts, we seem unable to react to their implications. Serious study of the effects of television on health and culture has only begun recently. Yet no drug in history has so quickly or completely isolated the entire culture of its users from contact with reality. And no drug in history has so completely succeeded in remaking in its own image the values of the culture that it has infected.

Television is by nature the dominator drug par excellence. Control of content, uniformity of content, repeatability of content make it inevitably a tool of coercion, brainwashing, and manipulation.[10] Television induces a trance state in the viewer that is the necessary precondition for brainwashing. As with all other drugs and technologies, television's basic character cannot be changed; television is no more reformable than is the technology that produces automatic assault rifles.

Television came along at precisely the right time from the point of view of the dominator elite. The nearly one hundred and fifty years of synthetic drug epidemics that began in 1806 had led to disgust at the spectacle of human degradation and spiritual cannibalism that institutional marketing of drugs created. In the same way that slavery eventually, when no longer convenient, became odious in the eyes of the very institutions that had created it, the abuse of drugs eventually triggered a backlash against this particular form of piratical capitalism. Hard drugs were made illegal. Of course underground markets then flourished. But drugs as stated instruments of national policy had been discredited. There would continue to be opium wars, instances of governments coercing other governments and peoples to produce or buy drugs—but in the future these wars would be dirty and secret, they would be "covert."

As the intelligence agencies that arose in the wake of World War II moved to take up their "deep cover" positions as the masterminds of the international narcotics cartels, the popular mind was turning on to television. Flattening, editing, and simplifying, television did its job and created a postwar American culture of the Ken-and-Barbie variety. The children of Ken and Barbie briefly broke out of the television intoxication in the mid-sixties through the use of hallucinogens. "Oops," responded the dominators, and they quickly made psychedelics illegal and halted all research. A double dose of TV therapy plus cocaine was ordered up for the errant hippies, and they were quickly cured and turned into consumption-oriented yuppies. Only a recalcitrant few escaped this leveling of values.[11] Nearly everyone learned to love Big Brother. And these few who don't are still clucked over by the dominator culture each time it compulsively scratches in the barnyard dust of its puzzlement over "what happened in the Sixties."

IV

PARADISE REGAINED?

14

A BRIEF HISTORY OF PSYCHEDELICS

Psychedelic plants and experience were first suppressed by European civilization, then ignored and forgotten. The fourth century witnessed the suppression of the mystery religions—the cults of Bacchus and Diana, of Attis and Cybele. The rich syncretism that was typical of the Hellenistic world had become a thing of the past. Christianity triumphed over the Gnostic sects—Valentinians, Marcionites, and others—which were the last bastions of paganism. These repressive episodes in the evolution of Western thought effectively closed the door on communication with the Gaian mind. Hierarchically imposed religion and, later, hierarchically dispensed scientific knowledge were substituted for any sort of direct experience of the mind behind nature.

The intoxicants of the Christian dominator culture, whether plants or synthetic drugs, were inevitably stimulants or narcotics—drugs of the workplace or drugs to dull care and pain. Drugs in the twentieth century serve only medical or recreational purposes. Yet even the West has retained the thin thread of remembrance of the Archaic, hierophantic, and ecstatic potential that certain plants hold.

The survival through long centuries in Europe of witchcraft and rites involving psychoactive plants attests that the gnosis of entering parallel dimensions by altering brain chemistry was never entirely lost. The plants of European witchcraft—thorn apple, mandrake, and nightshade—did not contain indole hallucinogens but were nevertheless capable of inducing intense altered states of consciousness. The Archaic connection of feminism to a magical dimension of risk and power was clearly perceived as a threat by the medieval church:

> As late as the Middle Ages the witch was still the *hagazussa*, a being that sat on the *Hag*, the fence, which passed behind the gardens and separated the village from the wilderness. She was a being who participated in both worlds. As we might say today, she was semi-demonic. In time, however, she lost her double features and evolved more and more into a representation of what was being expelled from culture, only to return, distorted, in the night.[1]

That these plants were the basis for entry into other dimensions was the result of the relative paucity in Europe of hallucinogen-containing species.

THE NEW WORLD HALLUCINOGENS

Indole-containing plant hallucinogens, and their cults, cluster in the tropical New World. The New World subtropical and tropical zones are phenomenally rich in hallucinogenic plants. Similar ecosystems in the Southeast Asian and Indonesian tropics cannot compare in numbers of endemic species that contain psychoactive indoles. Why are the Old World tropics, the tropics of Africa and Indonesia, not equally rich in hallucinogenic flora? No one has been able to answer this question. But statistically speaking the New World seems to be the preferred home of the more powerful psychoactive plants. Psilocybin, while now known to occur in European species, of diminutive mushrooms of the genus *Psilocybe*, has never been convincingly shown to have been a part of European shamanism or ethnomedicine. Yet its shamanic use in Oaxacan Mexico is three millennia old. Similarly, the New World has the only living cults based on use of dimethyltryptamine (DMT), the beta-carboline group including harmine, and the ergotlike complex in morning glories.

A historical consequence of this clustering of hallucinogens in the New World was that Western science discovered their existence rather late. This may explain the absence of "psychedelic" input into Western drugs for psychiatric uses. Meanwhile, because of the influence of hashish and opium on the Romantic imagination, the hashish reverie or opium dream became the paradigm of the action of the new "mental drugs" that fascinated the Bohemian literati from the late eighteenth century on. Indeed, hallucinogens were seen as capable of mimicking psychoses in their early encounter with Western psychotherapy.

In the nineteenth century explorer-naturalists began to return with more or less accurate ethnographic reports of the activities of aboriginal peoples. Botanists Richard Spruce and Alfred Russel Wallace traveled in the Amazon drainage in the 1850s. On the upper reaches of the Rio Negro, Spruce observed a group of Indians prepare an unfamiliar hallucinogen. He further observed that the main ingredient for this intoxicant was a liana, a woody climbing vine, which he named *Banisteria caapi*. Several years later, while traveling in western Ecuador he saw the same plant being used to make a hallucinogen called *ayahuasca*.[2] (See Figure 25.)

Ayahuasca has continued to the present day to be a part of the spiritual life of many of the tribes in the montane rain forest of South America. Immigrants into the Amazon basin have also accepted *ayahuasca* and have created their own ethnobotanical-medical system for using the psychedelic visions it imparts to promote healing.

The word *ayahuasca* is a Quechua word that roughly translates as "vine of the dead" or "vine of souls." This term refers not only to the prepared hallucinogenic beverage but also to one of its main ingredients, the woody liana. The tissues of this plant are rich in alkaloids of the beta-carboline type. The most important beta-carboline occurring in what is now called *Banisteriopsis caapi* is harmine. Harmine is an indole, but it is not overtly psychedelic unless taken in amounts that approach what is considered a toxic dose. However, well below that level, harmine is an effective short-acting monoamine oxidase inhibitor. Thus, a hallucinogen such as DMT,

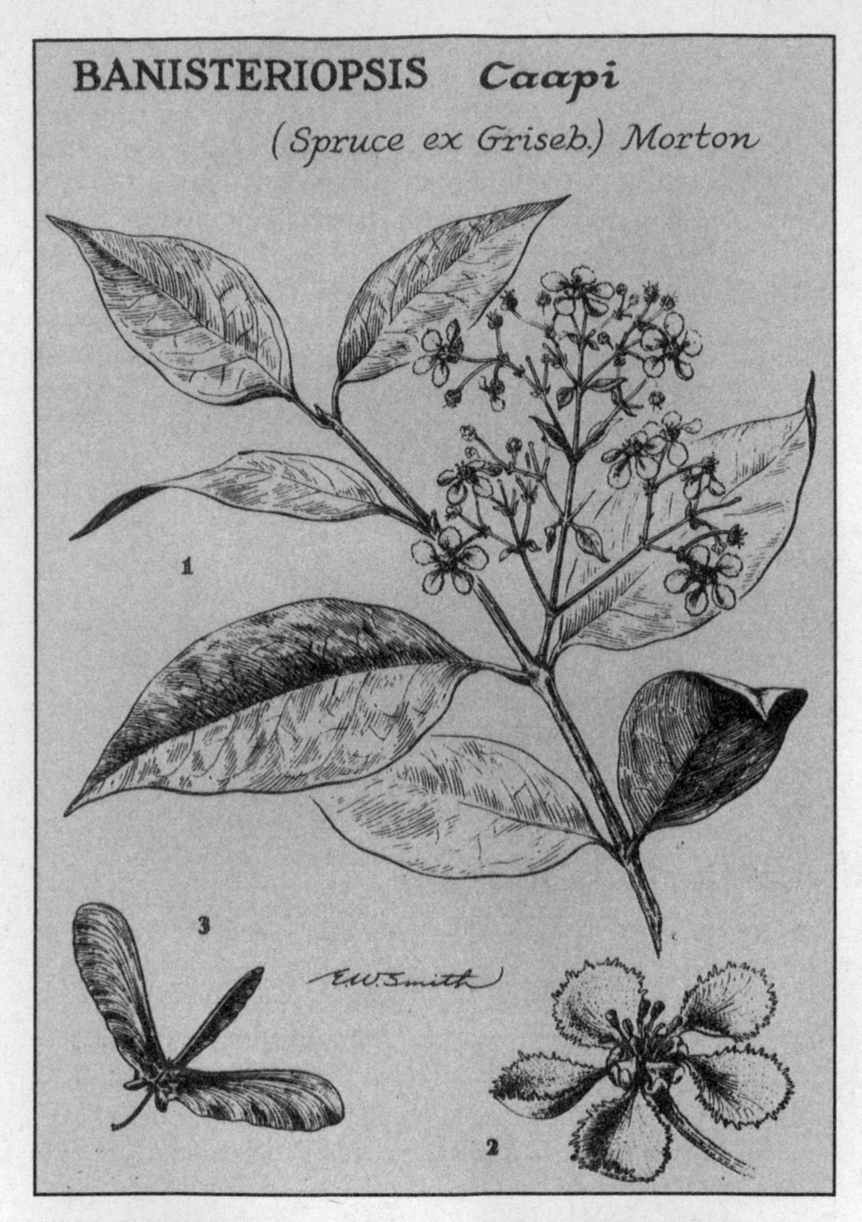

FIGURE 25. *Banisteriopsis caapi*, taxonomic drawing by E. W. Smith. From R. E. Schultes's *The Botany and Chemistry of Hallucinogens* (Springfield, MA: Charles Thomas, 1972), Fig. 27, p. 104.

which would normally be inactive if taken orally, is rendered highly psychoactive if taken orally in combination with harmine. Native peoples of the Amazon region have brilliantly exploited these facts in their search for techniques to access the magical dimensions crucial to shamanism.[3] By combining, in *ayahuasca*, DMT-containing plants with plants that contain MAO inhibitors, they have long exploited a pharmacological mechanism, MAO inhibition, not described by Western science until the 1950s.

In the presence of harmine, DMT becomes a highly psychoactive compound that enters the bloodstream and eventually makes its way past the blood-brain barrier and into the brain. There it very effectively competes with serotonin for synaptic bonding sites. This experience of the slow release of DMT lasts four to six hours and is the basis of the magical and shamanic view of reality that characterizes the *ayahuasquero* and his or her circle of initiates. Uninvolved or so-called objective styles of anthropological reportage have tended to underemphasize the culture-shaping importance that these altered states have had for tribal Amazonian societies.

The experience of ingesting *ayahuasca*—organic DMT taken in combination with the *Banisteriopsis* vine—has a number of characteristics that set it apart from the experience of smoking DMT. *Ayahuasca* is gentler and of much greater duration. Its themes and hallucinations are oriented toward the organic and the natural world, in marked contrast to the titanic, alien, and off-planet motifs that characterize the DMT flash. Why such major differences should exist between compounds that appear to be so structurally similar is an uninvestigated problem. Indeed, the whole relationship of particular kinds of visions to the compounds that elicit them is not well understood. In the native areas of its use, *ayahuasca* is regarded as a general-purpose healing elixir and is called *la purga*, the purge. Its effectiveness in combating intestinal parasites has been proven. Its effectiveness in killing the malaria organism is now being investigated. And its long history of effective shamanic use in folk psychiatry has been documented by Naranjo, Dobkin de Rios, Luna, and others.[4]

FIGURE 26. Tukano *ayahuasca* ritual in the Colombian Amazon. Courtesy of Fitz Hugh Ludlow Library.

AYAHUASCA

The experience induced by *ayahuasca* includes extremely rich tapestries of visual hallucination that are particularly susceptible to being "driven" and directed by sound, especially vocally produced sound. Consequently, one of the legacies of the *ayahuasca*-using cultures is a large repository of *icaros*, or magical songs (Figure 26). The effectiveness, sophistication, and dedication of an *ayahuasquero* is predicated upon how many magical songs he or she has effectively memorized. In the actual curing sessions, both patient and healer ingest *ayahuasca* and the singing of the magical songs is a shared experience that is largely visual.

The impact of long-term use of hallucinogenic indoles on mental and physical health is not yet well understood. My own experiences among the mestizo populations of Amazonas convince me that the long-term effect of *ayahuasca* use is an extraordinary state of health and integration. *Ayahuasqueros* use sound and suggestion to direct healing energy into parts of the body and unexamined aspects of an individual's personal history where psychic tension has come to rest. Often these methods exhibit startling parallels to the techniques of modern psychotherapy; at other times they seem to represent an understanding of possibilities and energies still unrecognized by Western theories of healing.

Most interesting from the point of view of the arguments made in this book are the persistent rumors of states of group-mind or telepathy that occur among the less acculturated tribal peoples. Our history of skepticism and empiricism would have us dismiss such claims as impossible, but we should think twice before doing so. The chief lesson to be learned from the psychedelic experience is the degree to which unexamined cultural values and limitations of language have made us the unwitting prisoners of our own assumptions. For it cannot be without reason that wherever in the world hallucinogenic indoles have been utilized, their use has been equated with magical self-healing and regeneration. The low incidence of serious mental illness among such populations is well documented.

THE FATHER OF PSYCHOPHARMACOLOGY

The modern era of psychopharmacology's interest in the aboriginal use of hallucinogenic plants has been extraordinarily brief. It dates to only a century ago, to German pharmacologist Lewis Lewin's tour of the United States.

On returning to Berlin in 1887, Lewin carried with him a quantity of peyote buttons, the vision-inducing cactus of the Sonoran Indians, that he had obtained from the Parke-Davis Company during his stay in Detroit. He set to work extracting, characterizing, and self-experimenting with the new compounds he discovered. Within a decade, peyote had attracted sufficient attention that in 1897 Philadelphia novelist and physician Silas Weir Mitchell became the first *gringo* to describe peyote intoxication:

> The display which for an enchanted two hours followed was such as I find it hopeless to describe in language which shall convey to others the beauty and splendor of what I

saw. Stars . . . delicate floating films of color . . . then an abrupt rush of countless points of white light swept across the field of view, as if the unseen millions of the Milky Way were to flow a sparkling river before the eyes . . . zigzag lines of very bright colors . . . the wonderful loveliness of swelling colors of more vivid colors gone before I could name them. Then, for the first time, definite objects associated with colors appeared. A white spear of grey stone grew up to huge height, and became a tall, richly finished Gothic tower of very elaborate and definite design, with many rather worn statues standing in the doorways or on stone brackets. As I gazed every projecting angle, cornice, and even the faces of the stones at their joining were by degrees covered or hung with clusters of what seemed to be huge precious stones, but uncut, some being like masses of transparent fruit.[5]

THE PLEASURES OF MESCALINE

In 1897, Arthur Heffter, a rival of Lewin's, became the first human being to isolate and ingest pure mescaline. Mescaline is a powerful visionary phenethylamine that occurs in the peyote cactus *Lophophora williamsii*. It has been used for at least several centuries by the Indians of Northern Mexico. Its use in Peru, where it is derived from species of cactus other than peyote, is at least several thousand years old.

The psychologist and pioneer sexologist Havelock Ellis, following the example of Weir Mitchell, soon offered his own account of mescaline's pleasures:

> The visions never resembled familiar objects; they were extremely definite, but yet always novel; they were constantly approaching, and yet constantly eluding, the semblance of known things. I would see thick, glorious fields of jewels, solitary or clustered, sometimes brilliant and sparkling, sometimes with a dull rich glow. Then they would spring up into flowerlike shapes beneath my gaze and then seem to turn into gorgeous butterfly forms or endless

> folds of glistening iridescent fibrous wings of wonderful insects. . . . Monstrous forms, fabulous landscapes, etc., appear. . . . It seems to us that any scheme which, in a detailed manner, assigns different kinds of visions to successive stages of the mescal state must be viewed as extremely arbitrary. The only thing that is typical with regard to sequence is that very elementary visions are followed by visions of a more complex character.[6]

Mescaline introduced experimenters to an agent of the *paradis artificiel* more potent than either cannabis or opium. Descriptions of mescaline states could hardly fail to attract the attention of the surrealists and psychologists who also shared a fascination with the images hidden in the depths of the newly defined unconscious. Dr. Kurt Beringer, Lewin's student and an acquaintance of Hermann Hesse's and Carl Jung's, became the father of psychedelic psychiatry. His phenomenological approach stressed reportage of the internal vistas beheld. He conducted hundreds of experiments with mescaline in human beings. The accounts given by his subjects are fascinating:

> Then the dark room once more. The visions of fantastic architecture again took hold of me, endless passages in Moorish style moving like waves alternated with astonishing pictures of curious figures. A design in the form of a cross was very frequent and present in unceasing variety. Incessantly the central lines of the ornament emanated, creeping like serpents or shooting forth like tongues toward the sides, but always in straight lines. Crystals appeared again and again, changing form and color and in the rapidity with which they came before my eyes. Then the pictures grew more steady, and slowly two immense cosmic systems were created, divided by a kind of line into an upper and a lower half. Shining with their own light, they appeared in unlimited space. From the interior new rays appeared in more luminescent colors, and gradually becoming perfect, they assumed the form of oblong prisms. At the same time they began to move. The systems approaching each other were attracted and repelled.[7]

In 1927 Beringer published his magnum opus *Der Meskalinrausch,* translated into Spanish but never into English. It is an inspired work, and it set the stage for the science of investigative pharmacology.

The following year saw the publication in English of Heinrich Klüver's *Mescal, the Divine Plant and Its Psychological Effects.* Klüver, whose work built on the observations of Weir Mitchell and Havelock Ellis, reintroduced the English-speaking world to the notion of visionary pharmacology. Especially important was the fact that Klüver took the hallucinogenic content of the experiences he was observing seriously and became the first to attempt to give a phenomenological description of the psychedelic experience:

> Clouds from left to right through optical field. Tail of a pheasant (in centre of field) turns into bright yellow star; star into sparks. Moving scintillating screw; "hundreds" of screws. A sequence of rapidly changing objects in agreeable colours. A rotating wheel (diameter about 1 cm.) in the centre of a silvery ground. Suddenly in the wheel a picture of God as represented in old Christian paintings.—Intention to see a homogeneous dark field of vision: red and green shoes appear. Most phenomena much nearer than reading distance.[8]

A MODERN RENAISSANCE

The investigation of hallucinogenic indoles also dates to the 1920s. A veritable Renaissance of psychopharmacology was taking place in Germany. In this atmosphere, Lewin and others became interested in harmine, an indole whose only source was thought to be *Banisteriopsis caapi,* the woody liana encountered by Richard Spruce nearly eighty years before. Indeed, Lewin's last published work reflected his new fascination with caapi; entitled *Banisteria Caapi, ein neues Rauschgift und Heilmittel,* it appeared in 1929. The excitement of Lewin and his colleagues was understandable: ethnographers such as the German Theodore Koch-Grünberg returned from Amazonas with accounts of tribes using telepathy-inducing plant drugs to direct the course of their societies. In 1927, the chemists E. Perrot and M. Raymond-Hamet isolated the active agent from *Banisteriopsis caapi* and named it telepathine. Years later, in 1957, researchers realized that telepathine was identical to the compound harmaline, extracted from *Peganum harmala*, and the name harmine was given official precedence over telepathine.

In the 1930s, the enthusiasm for the harmala alkaloids by and large vanished, as did much of the interest in ethnopharmacology. There were, however, notable exceptions. Among them was an Austrian expatriate living in Mexico.

Blas Pablo Reko, born Blasius Paul Reko, was a person of wide-ranging interests. His wandering life took him to the United States, to Ecuador, and finally to Oaxacan Mexico. There he became interested in ethnobotany and what is today called archaeo-astronomy, the study of ancient cultures' observations and attitudes toward the stars. Reko was an astute observer of the plant usages among the native people with whom he lived. In 1919, in rebuttal to an article by William Safford, Reko wrote that it was a hallucinogenic mushroom, and not peyote, that shamans of the Mixtec and Mazatecan people still used in a traditional way to induce visions.[9] In 1937, Reko sent Henry Wassén, an anthropologist and the curator of the ethnographical museum in Gothenburg, Sweden, a package containing collections of two plants that Reko had found particularly interesting. One of the samples was *piule* seed, the visionary morning glory seeds of *Ipomoea violacea*, which contain hallucinogenic indoles related to LSD.

Reko's other sample, unfortunately too decomposed to be identified to species, was a fragment of *teonanácatl*, the first specimen of a psilocybin-containing mushroom to be brought to scientific attention. Thus Reko initiated the study of the indole hallucinogens of Mexico and two chains of research and discovery, which would eventually be reunited when Albert Hofmann, the Swiss pharmaceutical chemist, characterized both compounds in his laboratory.

WHISPERS OF A NEW WORLD MUSHROOM

Reko had obtained his mushroom sample from Roberto Weitlander, a European engineer working in Mexico. The following year, 1938, a small group including Weitlander's daughter and anthropologist

Jean Basset Johnson became the first whites to attend a nightlong mushroom ceremony, or *velada*.

Wassén eventually forwarded Reko's samples to Harvard, where they came to the attention of the young ethnobotanist Richard Evans Schultes. Schultes had been a medical student until he had happened upon Klüver's work on mescaline. Schultes believed that Reko's mushroom might be the mysterious *teonanácatl* described by the Spanish chroniclers. He and an anthropology student from Yale, Weston La Barre, published a summation of the evidence for *teonanácatl* being a psychoactive mushroom.

The next year found Schultes accompanying Reko to the village of Huatla de Jiménez in the Sierra Mazatecan highlands. Specimens of psychoactive mushrooms were collected and forwarded to Harvard. But larger forces were afoot during the late thirties; like research in many other areas, ethnobotanical research slowed to a stop as the world slipped into world war. Reko retired, and as the Japanese solidified their hold on the rubber plantations of Malaya, Schultes accepted an assignment to the Amazon Basin to study rubber extraction for the U.S. government's wartime Office of Strategic Services. But, before this, in 1939, he published *The Identification of Teonanácatl, a Narcotic Basidiomycete of the Aztecs.*[10] Here he quietly announced his correct solution to an enigma that at that time seemed no more than a matter of scholarly debate among Mesoamericanists.

THE INVENTION OF LSD

Yet, even as the lights were going out in Europe, a fundamental breakthrough occurred. In 1938 Albert Hofmann was engaged in routine pharmaceutical research at Sandoz Laboratories, in Basel, Switzerland. Hofmann hoped to produce new drugs that would ease labor and childbirth. While working with the vasoconstricting substances derived from ergot, Hofmann synthesized the first d-lysergic acid diethylamide tartrate—LSD–25. Hofmann, a modest man, merely noted the correct completion of the synthesis, and the untested compound was cataloged and placed into storage. There it remained, surrounded by Nazi Europe for the next five years, five of the most tumultuous years in human history. It is frightening to imagine some of the possible consequences had Hofmann's discovery been recognized for what it was even a moment earlier.

Alfred Jarry may have anticipated and allegorized the great event when he wrote "The Passion Considered as an Uphill Bicycle Race"[11] in 1894. Indeed, the Dadaists and Surrealists and their forerunners grouped around Jarry and his *Ecole du Pataphysique* did much to explore the use of hashish and mescaline as augmentations to creative expression. They set the cultural stage for the truly surreal emergence of society's awareness of LSD. Every LSD enthusiast knows the story of how on April 16, 1943, feeling a touch of the Friday blahs, and unaware that he had absorbed a dose of LSD through handling the chemical without gloves, chemist, and soon-to-be counterculture hero, Albert Hofmann left work early and set off on his bicycle through the streets of Basel:

> I was forced to interrupt my work in the laboratory in the middle of the afternoon and proceed home, being affected by a remarkable restlessness, combined with a slight dizziness. At home I lay down and sank into a not unpleasant intoxicated dreamlike condition, characterized by an extremely stimulated imagination. In a dreamlike state, with eyes closed (I found the daylight to be unpleasantly glaring), I perceived an uninterrupted stream of fantastic pictures, extraordinary shapes with intense, kaleidoscopic play of colors. After some two hours this condition faded away.[12]

PANDORA'S BOX FLUNG OPEN

Finally, in 1947, the news of Hofmann's extraordinary discovery, a megahallucinogen active in the microgram range, surfaced in the scientific literature. As events in the 1950s made clear, Pandora's box had been flung open.

In 1954, Aldous Huxley wrote *The Doors of Perception*, a brilliant literary snapshot of the male European intellectual grappling with and agape at the realization of the true dimensions of consciousness and the cosmos:

> What the rest of us see only under the influence of mescaline, the artist is congenitally equipped to see all the time. His perception is not limited to what is biologically or socially useful. A little of the knowledge belonging to Mind at Large oozes past the reducing valve of brain and ego, into his consciousness. It is a knowledge of the intrinsic significance of every existent. For the artist, as for the mescaline taker, draperies are living hieroglyphs that stand in some peculiarly expressive way for the unfathomable mystery of our being. More even than the chair, though less perhaps than those wholly supernatural flowers, the folds of my gray flannel trousers were charged with "is-ness." To what they owed this privileged status, I cannot say.[13]

In 1956 the Czech chemist Steven Szara synthesized dimethyltryptamine, DMT. DMT remains the most powerful of all hallucinogens and one of the most short acting of these compounds known. When DMT is smoked, the intoxication reaches a peak in about two minutes and then abates over about ten minutes. Injections are typically more prolonged in their effect. Here is the discoverer's account:

> On the third or fourth minute after the injection vegetative symptoms appeared, such as tingling sensations, trembling, slight nausea, mydriasis, elevation of the blood pressure and increase of the pulse rate. At the same time eidetic phenomena, optical illusions, pseudo-hallucinations, and later real hallucinations appeared. The hallucinations consisted of moving, brilliantly colored oriental motifs, and later I saw wonderful scenes altering very rapidly.[14]

A year later, in May 1957, Valentina and Gordon Wasson published their now famous article in *Life* magazine announcing the discovery of the psilocybin mushroom complex. This article, as much as any other single piece of writing published on the subject, introduced into mass consciousness the notion that plants could cause exotic, perhaps even paranormal, visions. A New York investment banker, Wasson was well acquainted with the movers and shakers of the Establishment. Therefore, it was natural that he should turn to his friend Henry Luce, publisher of *Life*, when he needed a public forum in which to announce his discoveries. The tone of the *Life* article contrasts sharply with the hysteria and distortion that the American media would later fan. The article is both fair and detailed, both open-minded and scientific.

The chemical loose ends of the Wassons' discoveries were tidied up by Albert Hofmann, who made a second starring appearance in the history of psychedelic pharmacology by chemically isolating psilocybin and determining its structure in 1958.

In the short space of a dozen years in the recent past, from 1947 until 1960, the major indole hallucinogens were characterized, purified, and investigated. It is no coincidence that the subsequent decade was the most turbulent decade in America in a hundred years.

LSD AND THE PSYCHEDELIC SIXTIES

To understand the role of psychedelics in the 1960s, we must recall the lessons of prehistory and the importance to early human beings of the dissolution of boundaries in group ritual based on ingestion of hallucinogenic plants. The effect of these compounds is largely psychological and is only partially culturally conditioned; in fact, the compounds act to dissolve cultural conditioning of any sort. They force the corrosive process of reform of community values. Such compounds should be recognized as deconditioning agents; by revealing the relativity of conventional values, they become powerful forces in the political struggle to control the evolution of social images.

The sudden introduction of a powerful deconditioning agent such as LSD had the effect of creating a mass defection from community values, especially values based on a dominator hierarchy accustomed to suppressing consciousness and awareness.

LSD is unique among drugs in the power of its dose range. LSD is detectable in human beings at a dose of 50 micrograms, or 5/100,000 of a gram. Compounds that can elicit effects from amounts smaller than this are unheard of. This means that ten thousand doses of 100 micrograms each could in theory be obtained from one pure gram. More than any other aspect, this staggering ratio of physical mass to market value explains the meteoric rise of LSD use and its subsequent suppression. LSD is odorless and colorless, and it can be mixed in liquids; hundreds of doses could be concealed under a postage stamp. Prison walls were no barrier to LSD, nor were national borders. It could be manufactured in any location with the necessary technology and immediately transported anywhere. Millions of doses of LSD could be and were manufactured by a very few people. Pyramidal markets formed around these sources of supply; criminal syndicalism, a precondition to fascism, quickly followed.

But LSD is more than a commodity—it is a commodity that dissolves the social machinery through which it moves. This effect has bedeviled all the factions that have sought to use LSD to advance a political agenda.

A psychological deconditioning agent is inherently counteragenda. Once the various parties attempting to gain control of the situation recognized this, they were able to agree on one thing—that LSD be stopped. How and by whom this was done is a lively story that has been well told, most notably by Jay Stevens in *Storming Heaven* and Martin Lee and Bruce Shlain in *Acid Dreams*.[15] These authors make clear that when the methods that worked for colonial empires peddling opium in the nineteenth century were applied by the CIA to the internal management of the American state of mind during the Vietnam War they damn near blew up the whole psychosocial shithouse.

Lee and Shlain write:

> The use of LSD among young people in the US reached a peak in the late 1960s, shortly after the CIA initiated a series of covert operations designed to disrupt, discredit, and neutralize the New Left. Was this merely a historical coincidence, or did the Agency actually take steps to promote the illicit acid trade? Not surprisingly, CIA spokesmen dismiss such a notion out of hand. "We do not target American citizens," former CIA director Richard Helms told the American Society of Newspaper Editors in 1971. "The nation must to a degree take it on faith that we who lead the CIA are honorable men, devoted to the nation's service."

> Helms's reassurances are hardly comforting in light of his own role as the prime instigator of Operation MK-ULTRA, which utilized unwitting Americans as guinea pigs for testing LSD and other mind-altering substances.
>
> As it turns out, nearly every drug that appeared on the black market during the 1960s—marijuana, cocaine, heroin, PCP, amyl nitrate, mushrooms, DMT, barbiturates, laughing gas, speed, and many others—had previously been scrutinized, tested, and in some cases refined by CIA and army scientists. But of all the techniques explored by the Agency in its multimillion-dollar twenty-five-year quest to conquer the human mind, none received as much attention or was embraced with such enthusiasm as LSD-25. For a time CIA personnel were completely infatuated with the hallucinogen. Those who first tested LSD in the early 1950s were convinced that it would revolutionize the cloak and dagger trade. During Helms's tenure as CIA director, the Agency conducted a massive illegal domestic campaign against the antiwar movement and other dissident elements in the US.[16]

As a result of Helms' successful campaign, the New Left was in a shambles when Helms retired from the CIA in 1973. Most of the official records pertaining to the CIA's drug and mind control projects were summarily destroyed on orders from Helms shortly before his departure. The files were shredded, according to Dr. Sidney Gottlieb, chief of the CIA's Technical Services Staff, because of "a burgeoning paper problem." Lost in the process were numerous documents concerning the operational employment of hallucinogenic drugs, including all existing copies of a classified CIA manual titled "LSD: Some Un-Psychedelic Implications."[17]

The times were extraordinary, made only more so by the fantasies of those who sought to control them. The 1960s can almost be seen as a time when two pharmacological mind-sets clashed in an atmosphere close to that of war. On the one hand, international heroin syndicates sought to narcotize America's black ghettos, while hoodwinking the middle class into supporting military adventurism. On the other, self-organized criminal syndicates manufactured and distributed tens of millions of doses of LSD while waging a highly visible underground campaign for their own brand of psychedelic cryptoanarchy.

The result of this encounter can be seen as something of a standoff. The war in Southeast Asia was a catastrophic defeat for the American Establishment, yet paradoxically barely a shred of psychedelic utopianism survived the encounter. All psychedelic drugs, even such unknowns as ibogaine and bufotinin, were made illegal. A relentless restructuring of values was begun in the West; throughout the seventies and eighties the need to deny the impact of the sixties took on something of the flavor of a mass obsession. As the seventies progressed, the new management agenda became clear; while heroin had lost some of its glamour, now there was to be television for the poor and cocaine for the rich.

By the end of the 1960s psychedelic research had been hounded out of existence—not only in the United States, but around the world. And this happened despite the enormous excitement these discoveries had created among psychologists and students of human behavior, an excitement analogous to the feelings that swept the physics community at the news of the splitting of the atom. But whereas the power of the atom, convertible into weapons of mass destruction, was fascinating to the dominator Establishment, the psychedelic experience loomed ultimately as an abyss.

The new era of repression came despite the fact that a number of researchers were using LSD to cure conditions previously considered untreatable. Canadian psychiatrists Abram Hoffer and Humphrey Osmond tabulated the results of eleven separate studies of alcoholism and concluded that 45 percent of the patients treated with LSD improved.[18] Promising results were obtained in attempts to treat schizophrenics, autistic children, and the severely depressed. Many of these findings were attacked after LSD became illegal, but better experiments were never designed and the work could not be repeated because of its illegality. Psychiatry's promising new uses of LSD to treat pain, addiction, alcoholism, and depression during terminal illness were put on indefinite hold.[19] It fell to the humble science of botany to advance our understanding of hallucinogenic plants.

RICHARD SCHULTES AND THE PLANT HALLUCINOGENS

At the center of this quiet revolution in botany was a single man, Richard Evans Schultes—the same Schultes who had seen his Mexican research interrupted by World War II. Schultes spent more than fifteen years in the Amazon Basin; he filed reports with the OSS on the natural rubber crop until the invention of synthetic rubber made that task unnecessary; and he studied and collected the orchids of the rain forest and the *alti plano.* As Schultes traveled, it became clear that his interest in Klüver's experiments with mescaline, and his fascination with the psychoactive plants of Mexico, would not be wasted in South America.

Years later, he would write of his work among the shamans of the Sibundoy Valley of southern Colombia: "The shamanism of this valley may well represent the most highly evolved narcotic consciousness on earth." What was true of the Sibundoy was nearly as true of the Upper Amazon generally, and over the next several decades it was Schultes and his graduate students who practiced and spread the gospel of modern ethnobotany.

Schultes focused on psychoactive plants from the beginning of his work. He correctly recognized that aboriginal people who had painstakingly composed an armamentorium of healing and medicinal plants were likely to most clearly understand their mental effects. After his early work on peyote and mushrooms, Schultes turned his attention to the several species of vision-inducing morning glories used in Oaxaca. In 1954 he published on the snuffs of the Amazon and thus announced to the world the existence of traditional shamanic usage of plant-produced DMT.

Throughout the next thirty-five years the Harvard group meticulously investigated and published all instances of psychoactive plant usage that came to their attention. This body of now continuously expanding work—an integrated body of taxonomic, ethnographic, pharmacological, and medical information—constitutes the core of the data base currently in global use.

The birth of ethnopsychopharmacology took place at Harvard under Schultes's watchful eye, much of it during the turbulent years when Timothy Leary was also at Harvard and attracting a very different sort of reputation through his own effort to place the psychedelic experience on the social agenda.

LEARY AT HARVARD

It is doubtful that either Leary or Schultes saw much to like in the other. They could hardly have been more different—Schultes the reticent Brahmin, scholar, and botanist/scientist, Leary the shamanic trickster and social scientist. Leary's earliest psychedelic experience had been with mushrooms; he would later recall that he was recruited for what he called "my planetary mission" by that first psilocybin encounter in Mexico. But the politics of expediency were forced on the Harvard Psilocybin Project; LSD was more accessible and less expensive than psilocybin. Michael Hollingshead was the person most responsible for making LSD the drug of choice in Harvard's psychedelic circles:

> [Leary] latched onto Hollingshead as his guru. Leary followed him around for days on end. . . . Richard Alpert and Ralph Metzner, two of Leary's closest associates, were vexed to see him in such a helpless state. They thought he had really blown his mind and they blamed Hollingshead. But it was only a matter of time before they too sampled the contents of the mayonnaise jar. Hollingshead gave the drug to the members of the psilocybin project and from then on LSD was part of their research repertoire.[20]

PSILOCYBIN: PSYCHEDELICS IN THE SEVENTIES

After the suppression of the psychedelic subculture that began with the illegalization of LSD in October 1966, the evolution of substance sophistication seemed to lose momentum. The most significant development during the 1970s from the point of view of those alerted to the psychedelic potential by earlier experiences with LSD and mescaline was the appearance, beginning in late 1975, of techniques and manuals for the home cultivation of psilocybin mushrooms. Several such manuals appeared, the earliest being *Psilocybin: The Magic Mushroom Grower's Guide* written by my brother and me and published pseudonymously under the names O. T. Oss and O. N. Oeric. The book sold over a hundred thousand copies over the next five years, and several imitators also did very well. Hence psilocybin, long sought and long familiar to the psychedelic community through the effusive prose of Wasson and Leary, became available at last to large numbers of people, who no longer needed to travel to Oaxaca to obtain the experience.

The ambience of psilocybin is different from that of LSD. Hallucinations come easier, and so does a sense that this is not merely a lens for the inspection of the personal psyche, but a communication device for getting in touch with the world of the high shamanism of Archaic antiquity. A community of therapists and astronauts of inner space has evolved around the use of the mushrooms. To this day these quiet groups of professionals and inner pioneers constitute the core of the community of people who have admitted the fact of the psychedelic experience into their lives and professions and who continue to grapple with it and learn from it.

And there we will leave the history of human involvement with plants that intoxicate or bring visions or consuming frenzy. We now know no more really than was known by our remote ancestors. Perhaps less. Indeed, we cannot even be certain whether science, the epistemic tool upon which we have come to depend most heavily, is up to this task. For we can begin our quest for understanding in the cool domains of archaeology or botany or neuropharmacology, but what is troubling and miraculous is the fact that all these approaches, when seen with psychedelic eyes, seem to lead to the internal nexus of self and world that we experience as the deepest levels of our own being.

PSYCHEDELIC IMPLICATIONS

What does it mean that pharmacology's effort to reduce the mind to molecular machinery confined within the brain has handed back to us a vision of mind that argues for its almost cosmic proportions? Drugs seem the potential agents of both our devolution back into the animal and our metamorphosis into a shining dream of possible perfection. "Man to man is like unto an errant beast," wrote the English social philosopher Thomas Hobbes, "and man to man is like unto a god." To this we might add "and never more so than when using drugs."

The 1980s were an era unusually empty of developments in the area of psychedelics. Synthetic amphetamines such as MDA were sporadically available from the early 1970s on, and during the 1980s MDMA (Ecstasy) appeared in significant amounts. MDMA in particular showed promise when used with directed psychotherapy,[21] but these drugs were quickly made illegal and forced underground before they achieved any general impact on society. MDMA was simply the most recent echo of the search for inner balance that drives ever-shifting styles of drug use and inner exploration. The drug terror of the 1980s was crack cocaine, a drug whose economic profile and high risk for addiction made it ideal in the eyes of the already established infrastructure for supplying the ordinary cocaine market.

The costs of drug education and drug treatment are small relative to routine military expenditures and could be contained. What cannot be contained are the effects that psychedelics would have in shaping the cultural self-image if all drugs were legal and available. This is the hidden issue that makes governments unwilling to consider legalization: the unmanaged shift of consciousness that legal and available drugs, including plant psychedelics, would bring is extremely threatening to a dominator, ego-oriented culture.

PUBLIC AWARENESS OF THE PROBLEM

To this point public awareness of issues concerning drugs has been lacking and public opinion easily manipulated. The situation must change. We must prepare to master the problem of our relationship to psychoactive substances. This cannot be done by an appeal to some antihuman standard of behavior that spells more suppression of the mass psyche by dominator metaphors. There can be no "Saying No" to drugs; nothing so asinine or preposterous will do. Nor can we be led down the primrose path by feel-good philosophies that see unbridled hedonism as the Holy Grail of social organization. Our only reasonable course is decriminalization of drugs, mass education, and shamanism as an interdisciplinary and professional approach to these realities. It is our souls that have become ill when we abuse drugs, and the shaman is a healer of souls. Such measures will not immediately solve the general drug problem, but they will preserve the sorely needed pipeline to the spirit that we must have if we expect to restructure society's attitude toward plant and substance use and abuse.

An interrupted psychophysical symbiosis between ourselves and the visionary plants is the unrecognized cause of the alienation of modernity and the cultural mind-set of planetary civilization. A worldwide attitude of fear toward drugs is being fostered and manipulated by the dominator culture and its propaganda organs. Vast illicit fortunes continue to be made; government continues to wring its hands. This is but the most recent effort to profiteer from and frustrate our species' deeply instinctual need to make contact with the Gaian mind of the living planet.

15
ANTICIPATING THE ARCHAIC PARADISE

Let us turn to the kind of options available to someone who seriously wishes to redress the history-created ego imbalance within themselves. This requires a brief survey of the opportunities to explore plant hallucinogens presently afforded by non-Western societies around the world.

REAL WORLD OPTIONS

There is, of course, the psilocybin complex discovered by Valentina and Gordon Wasson—the magic mushrooms of central Mexico, which almost certainly played a major role in the religion of the Mayan and Toltec civilizations. This complex includes the more widely distributed *Stropharia cubensis*, which is thought to have originated in Thailand but is now found throughout the warm tropics.

The highlands of Mazetecan Mexico are home to two species of morning glories. *Ipomoea purpura* and *Turbina* (formerly *Rivea*) *corymbosa*. The properties of ergot that interested Albert Hofmann and led eventually to his discovery of LSD, that of being a constrictor of smooth muscle and thus a potential aid in labor, had long been known to midwives of the Sierra Mazateca. The accompanying dissolution of perceived boundaries and influx of visionary information made these morning glories the preferred substitute in times when psilocybin-containing mushrooms were not available.[1]

With only one exception, all of the shamanic vision plants—including the morning glory complex of Mexico and the psilocybin complex—turn out to be hallucinogenic indoles. The single exception is mescaline, which is a kind of phenethylamine.

And one must not fail to consider those other indoles, the short-acting tryptamines and the beta-carbolines. The short-acting tryptamines can be used separately or in combination with beta-carbolines. The beta-carbolines, though hallucinogenic in themselves, are most effective when used as monoamine oxidase inhibitors to enhance the effects of short-acting tryptamines and also to cause tryptamines to become orally active.

I have not mentioned any synthetics, because I would prefer to separate the vision-producing plants from the popular notion of drugs. The global drug problem is a different issue entirely and has to do with the fates of nations and mega-dollar criminal syndicates. I avoid synthetic drugs and prefer the organic hallucinogens, because I believe that a long history of shamanic usage is the first seal of approval that one must look for when selecting a substance for its possible effects on personal growth. And if a plant has been used for thousands of years, one can also be fairly confident that it does not cause tumors or miscarriages or carry other unacceptable physical risks. Over time, trial and error has resulted in the choice of the most effective and least toxic plants for shamanic use.

Other criteria are also relevant when evaluating a substance. It is important to use only those compounds that do not insult the physical brain; regardless of what the physical brain does or doesn't have to do with the mind, it certainly has much to do with the metabolism of hallucinogens. Compounds alien to the brain and therefore difficult for it to metabolize should be avoided.

One way of judging how long a relationship between humans and a plant has been in place is to notice how benign the compound is in human metabolism. If after you have taken a plant, your eyes are not in focus forty-eight hours later, or your knees are feeling rubbery three days later, then this is not a benign compound that has evolved into a smooth hand-in-glove fit with the human user.

THE CASE FOR HALLUCINOGENIC TRYPTAMINES

These criteria explain why, to my mind, the tryptamines are so interesting, and why I argue for the psilocybin mushroom as the primary hallucinogen involved in the Archaic origin of consciousness. The tryptamines, including psilocybin, bear a striking resemblance to human neurochemistry. The human brain, and indeed most nervous systems, run partially on 5-hydroxytryptamine, also known as serotonin. DMT, closely related to serotonin, is the hallucinogenic compound central to Amazonian shamanism, and is the most powerful of all hallucinogens in human beings and yet when smoked clears the system in less than fifteen minutes. The structural similarity between these two compounds may indicate the great antiquity of the evolutionary relationship between human brain metabolism and these particular compounds.

Having discussed options, it only remains to discuss techniques. Aldous Huxley called the psychedelic experience "a gratuitous grace." By this he meant that by itself the psychedelic experience is neither necessary nor sufficient for personal salvation. It also can be elusive. All conditions for success may be present and one can still fail to connect. However, one cannot fail to connect if all conditions for success are present and one does it over and over again—perhaps there is a temporal variable there.

Good technique is obvious: one sits down, one shuts up, and one pays attention. That is the essence of good technique. These journeys should be taken on an empty stomach, in silent darkness, and in a situation of comfort, familiarity, and security. "Set" and "setting," terms established by Timothy Leary and Ralph Metzner in the 1960s, have remained excellent reference points.[2] Set refers to the interiorized feelings, hopes, fears, and expectations of the would-be psychonaut. Setting refers to the external situation in which the interior journey will take place—the noise level, light level, and level of familiarity to the voyager. Both set and setting should optimize feelings of security and confidence. External stimuli should be severely limited—phones unplugged, noisy machines stilled. Study the darkness behind closed eyelids with the expectation of seeing something. The experience is not simply eidetic hallucination (which we get when we press on our closed eyelids), although it begins like eidetic hallucination. Comfortable, silent darkness is the preferred environment for the shaman to launch what the neo-Platonic mystic Plotinus called "the flight of the alone to the Alone."

Major conceptual and linguistic difficulties are involved in conveying to people precisely what this experience is like. Most of those reading my words will have had at some point in their lives something which they would describe as a "drug experience." But did you know that your experience is bound to be unique and different from that of everyone else? These experiences range from mild tingling in the feet to being in titanic and alien realms where the mind boggles and language fails. And one feels the presence of the utterly unspeakable, the wholly Other. Memories fall, gritty and particulate, like the snows of yesteryear. Opalescence anticipates neon, and language gives birth to itself. Hyperbole becomes impossible. And therein lies the importance of discussing these matters.

HOW DOES IT FEEL?

What was the ambience of that lost Edenic world? What is the feeling whose absence has left us stranded in history? The onset of an indole hallucinogen is characterized first by a somatic activation, a feeling in the body. The indoles are not soporifics but central nervous system stimulants. The familiar feeling of "fight or flight" is often a feature of the first wave of somatic feelings associated with the hallucinogen. One must discipline the hind brain and simply wait through this turmoil within the animal body.

An orally active compound such as psilocybin makes its full effects felt in about an hour and a half; a compound that is smoked, such as DMT, becomes active in less than a minute. By whatever route the indole hallucinations are triggered, their full unfolding is impressive indeed. Bizarre ideas, often hilariously funny, curious insights, some seeming almost godlike in their profundity, shards of memories and free-form hallucinations all clamor for attention. In the state of hallucinogenic intoxication, creativity is not something that one expresses; it is something that one observes.

The existence of this dimension of knowable meaning that appears to be without connection to one's personal past or aspirations seems to argue that we are facing either a thinking Other or the deep structures of the psyche made suddenly visible. Perhaps both. The profundity of this state and its potential for a positive feedback into the process of reorganizing the personality should have long ago made psychedelics an indispensable tool for psychotherapy. After all, dreams have made a major claim on the attention of the theoreticians of psychic process, as have free association and hypnotic regression; yet these are but peepholes into the hidden world of psychic dynamics compared with the expansive view that psychedelics provide.

FACING THE ANSWER

The situation that we now must deal with is not one of seeking the answer, but of facing the answer. The answer has been found; it just happens to lie on the wrong side of the fence of social toleration and legality. We are thus forced into a strange little dance. Those professionally involved know that psychedelics are the most powerful instruments for the study of the mind that are possible to conceive. And yet these people often work in academia and must frantically try to ignore the fact that the answer has been placed in our hands. Our situation is not unlike that of the sixteenth century when the telescope was invented and shattered the established paradigm of the heavens. The 1960s proved that we are not wise enough to take the psychedelic tools into our hands without a social and intellectual transformation. This transformation must begin now with each of us.

Nature, in her evolutionary and morphogenetic richness, has offered a compelling model for us to follow in the shamanic task of re-sacralization and self-transformation that lies ahead. The totemic animal image for the future human to model is the octopus. This is because the cephalopods, the squids and octopi, lowly creatures though they may seem, have perfected a form of communication that is both psychedelic and telepathic—an inspiring model for the human communications of the future.

CONSIDER THE OCTOPUS

An octopus does not communicate with small mouth noises, even though water is a good medium for acoustic signaling. Rather, the octopus becomes its own linguistic intent. Octopi have a large repertoire of color changes, dots, blushes, and traveling bars that move across their surfaces. This repertoire in combination with the soft-bodied physique of the creature allows it to obscure and reveal its linguistic intent simply by rapidly folding and unfolding the changing parts of its body. The mind and the body of the octopus are the same and hence equally visible; the octopus wears its language like a kind of second skin. Octopi can hardly not communicate. Indeed, their use of "ink" clouds to conceal themselves may indicate that this is the only way that they can have anything like a private thought. The ink cloud may be a kind of correction fluid for voluble octopi who have misstated themselves. Martin Moyniham has written of the complexities of cephalopod communication:

> The communication and related systems of . . . cephalopods are largely visual. They include arrangements of pigment cells, postures, and movements. The postures and movements can be ritualized or unritualized. Color changes presumably are always ritualized. The various patterns can be combined in many and often intricate ways. They can be changed very rapidly. Since they are visual, they should be relatively easy to describe and to decipher by human observers. There are, however, complications. . . .
>
> Read or not, correctly or not, the patterns of cephalopods, like those of all other animals, encode information. When and insofar as they are messages, intentional or not, [they] would seem to have not only syntax but also a simple grammar.[3]

Like the octopi, our destiny is to become what we think, to have our thoughts become our bodies and our bodies become our thoughts. This is the essence of the more perfect Logos envisioned by the Hellenistic polymath Philo Judaeus—a Logos, an indwelling of the Goddess, not heard but beheld. Hans Jonas explains Philo Judaeus's concept as follows:

A more perfect archetypal logos, exempt from the human duality of sign and thing, and therefore not bound by the forms of speech, would not require the mediation of hearing, but is immediately beheld by the mind as the truth of things. In other words the antithesis of seeing and hearing argued by Philo lies as a whole within the realm of "seeing"—that is to say, it is no real antithesis but a difference of degree relative to the ideal of immediate intuitive presence of the object. It is with a view to this ideal that the "hearing" here opposed to "seeing" is conceived, namely as its deputizing, provisional mode, and not as something authentic, basically other than seeing. Accordingly the turn from hearing to seeing here envisaged is merely a progress from a limited knowledge to an adequate knowledge of the same and within the same project of knowledge.[4]

ART AND THE REVOLUTION

The Archaic Revival is a clarion call to recover our birthright, however uncomfortable that may make us. It is a call to realize that life lived in the absence of the psychedelic experience upon which primordial shamanism is based is life trivialized, life denied, life enslaved to the ego and its fear of dissolution in the mysterious matrix of feeling that is all around us. It is in the Archaic Revival that our transcendence of the historical dilemma actually lies.

There is something more. It is now clear that new developments in many areas—including mind-machine interfacing, pharmacology of the synthetic variety, and data storage, imaging, and retrieval techniques—are coalescing into the potential for a truly demonic or an angelic self-imaging of our culture. Those who are on the demonic side of this process are fully aware of this potential and are hurrying full tilt forward with their plans to capture the technological high ground. It is a position from which they hope to turn nearly everyone into a believing consumer in a beige fascism from whose image factory none will escape.

The shamanic response, the Archaic response, the human response, to this situation should be to locate the art pedal and push it to the floor. This is one of the primary functions of shamanism, and is the function that is tremendously synergized by the psychedelics. If psychedelics are exopheromones that dissolve the dominant ego, then they are also enzymes that synergize the human imagination and empower language. They cause us to connect and reconnect the contents of the collective mind in ever more implausible, beautiful, and self-fulfilling ways.

If we are serious about an Archaic Revival, then we need a new paradigmatic image that can take us rapidly forward and through the historical choke point that we can feel impeding and resisting a more expansive, more humane, more caring dimension that is insisting on being born. Our sense of political obligation, of the need to reform or save the collective soul of humanity, our wish to connect the end of history with the beginning of history—all of this should impel us to look at shamanism as an exemplary model. In the current global crisis we cannot fail to take its techniques seriously, even those which may challenge the divinely ordained covenants of the constabulary.

CONSCIOUSNESS EXPANSION

Years ago, before Humphrey Osmond coined the term "psychedelic," there was current a phenomenological description for psychedelics; they were called "consciousness-expanding drugs." I believe that this is a very good description. Consider our dilemma on this planet. If the expansion of consciousness does not loom large in the human future, what kind of future is it going to be? To my mind, the propsychedelic position is most fundamentally threatening to the Establishment because, when fully and logically thought through, it is an antidrug, antiaddiction position. And make no mistake about it; the issue is drugs. How drugged shall you be? Or, to put it another way, how conscious shall you be? *Who* shall be conscious? *Who* shall be unconscious?

We need a serviceable definition of what we mean by "drug." A drug is something that causes unexamined, obsessive, and habitual behavior. You don't examine obsessive behavior; you just do it. You let nothing get in the way of your gratification. This is the kind of life that we are being sold at every level. To watch, to consume, and to watch and consume yet more. The psychedelic option is off in a tiny corner, never mentioned; yet it represents the only counterflow directed against a tendency to leave people in designer states of consciousness. Not their own designs, but the designs of Madison Avenue, of the Pentagon, of the Fortune 500 corporations. This isn't just metaphor; it is really happening to us.

Looking down on Los Angeles from an airliner, I never fail to notice that it is like looking at a printed circuit: all those curved driveways and cul de sacs with the same little modules installed along each one. As long as the *Reader's Digest* stays subscribed to and the TV stays on, these modules are all interchangeable parts within a very large machine. This is the nightmarish reality that Marshall McLuhan and Wyndham Lewis and others foresaw: the creation of the public as herd. The public has no history and no future, the public lives in a golden moment created by a credit system which binds them ineluctably to a web of illusions that is never critiqued. This is the ultimate consequence of having broken off the symbiotic relationship with the Gaian matrix of the planet. This is the consequence of lack of partnership; this is the legacy of imbalance between the sexes; this is the terminal phase of a long descent into meaninglessness and toxic existential confusion.

The credit for giving us tools to resist this horror belongs to unsung heroes who are botanists and chemists, people such as Richard Schultes, the Wassons, and Albert Hofmann. Thanks to them we have, in this most chaotic of centuries, taken into our frail hands the means to do something about our predicament. Psychology, in contrast, has been complacent and silent. Psychologists have been content with behaviorist theory-making for fifty years, while knowing in their hearts that they were doing a potentially fatal disservice to human dignity, by ignoring the potential of psychedelics.

THE DRUG WAR

If there was ever a moment to be heard and be counted and to try to clarify thinking on these issues, that moment is now. For some time there has been a major attack on the Bill of Rights under the pretext of the so-called drug war. Somehow the drug issue is even more frightening to the public herd than was Communism, even more insidious.

The quality of rhetoric emanating from the psychedelic community must improve radically. If it does not, we will forfeit the reclamation of our birthright and all opportunity for exploring the psychedelic dimension will be closed off. Ironically, this tragedy could occur almost as a footnote to the suppression of synthetic and addictive narcotics. It cannot be said too often: the psychedelic issue is a civil rights and civil liberties issue. It is an issue concerned with the most basic of human freedoms: religious practice and the privacy of the individual mind.

It was said that women could not be given the vote because society would be destroyed. Before that, kings could not give up absolute power because chaos would result. And now we are told that drugs cannot be legalized because society would disintegrate. This is puerile nonsense! As we have seen, human history could be written as a series of relationships with plants, relationships made and broken. We have explored a number of ways in which plants, drugs, and politics have cruelly intermingled—from the influence of sugar on mercantilism to the influence of coffee on the modern office worker, from the British forcing opium on the Chinese population to the CIA using heroin in the ghetto to choke off dissent and dissatisfaction.

History is the story of these plant relationships. The lessons to be learned can be raised into consciousness, integrated into social policy, and used to create a more caring, meaningful world, or they can be denied just as discussion of human sexuality was repressed until the work of Freud and others brought it into the light. The analogy is apt because the enhanced capacity for cognitive experience made possible by plant hallucinogens is as basic a part of our humanness as is our sexuality. The question of how quickly we develop into a mature community able to address these issues lies entirely with us.

HYPERSPACE AND HUMAN FREEDOM

What is most feared by those who advocate the unworkable Luddite solution of "Just say no" is a world in which all traditional community values have dissolved in the face of an endless search for self-gratification on the part of drug-obsessed individuals and populations. We should not dismiss this only too real possibility. But what must be rejected is the notion that this admittedly disturbing future can be avoided by witch hunts, the suppression of research, and the hysterical spreading of disinformation and lies.

Drugs have been a part of the galaxy of cultural concerns since the dawn of time. It was only with the advent of technologies capable of refining and concentrating the active principles of plants and plant preparations that drugs separate themselves from the general background of cultural concerns and become instead a scourge.

In a sense what we have is not a drug problem, but a problem with the management of our technologies. Is our future to include the appearance of new synthetic drugs, a hundred or a thousand times more addictive than heroin or crack? The answer is absolutely yes—unless we bring to consciousness and examine the basic human need for chemical dependency and then find and sanction avenues for expression of this need. We are discovering that human beings are creatures of chemical habit with the same horrified disbelief as when the Victorians discovered that humans are creatures of sexual fantasy and obsession. This process of facing ourselves as a species is a necessary precondition to the creation of a more humane social and natural order. It is important to remember that the adventure of facing who we are did not begin or end with Freud and Jung. The argument this book has sought to develop is that the next step in the adventure of self-understanding can begin only when we take note of our innate and legitimate need for an environment rich in mental states that are induced through an act of will. I believe we can initiate the process by revisioning our origins. Indeed, I have taken great pains to show that in the Archaic milieu in which self-reflection first emerged we find clues to the roots of our own troubled history.

WHAT IS NEW HERE

The hallucinogenic indoles, unstudied and legally suppressed, are here presented as agents of evolutionary change. They are biochemical agents whose ultimate impact is not on the direct experience of the individual but on the genetic constitution of the species. Earlier chapters drew attention to the fact that increased visual acuity, increased reproductive success, and increased stimulation of protolinguistic brain functions are all logical consequences of the inclusion of psilocybin in the early human diet. If the notion that human consciousness emerged out of indole-mediated synergy of neurodevelopment could be proven, then our image of ourselves, our relationship to nature, and the present dilemma over drug use in society would change.

There is no solution to the "drug problem," or to the problem of environmental destruction or the problem of nuclear weapons stockpiles, until and unless our self-image as a species is reconnected to the earth. This begins with an analysis of the unique confluence of conditions that must have been necessary for animal organization to make the leap to conscious self-reflection in the first place. Once the centrality of the hallucinogen-mediated human-plant symbiosis in the scenario of our origins is understood, we are then in a position to appreciate our current state of neurosis. Assimilation of the lessons contained in those ancient and formative events can lay the groundwork for solutions to meet not only society's need to manage substance use and abuse but also our deep and growing need for a spiritual dimension to our lives.

THE DMT EXPERIENCE

Earlier in this chapter, DMT was singled out as being of particular interest. What can be said of DMT as an experience and in relation to our own spiritual emptiness? Does it offer us answers? Do the short-acting tryptamines offer an analogy to the ecstasy of the partnership society before Eden became a memory? And if they do, then what can we say about it?

What has impressed me repeatedly during my many glimpses into the world of the hallucinogenic indoles, and what seems generally to have escaped comment, is the transformation of narrative and language. The experience that engulfs one's entire being as one slips beneath the surface of the DMT ecstasy feels like the penetration of a membrane. The mind and the self literally unfold before one's eyes. There is a sense that one is made new, yet unchanged, as if one were made of gold and had just been recast in the furnace of one's birth. Breathing is normal, heartbeat steady, the mind clear and observing. But what of the world? What of incoming sensory data?

Under the influence of DMT, the world becomes an Arabian labyrinth, a palace, a more than possible Martian jewel, vast with motifs that flood the gaping mind with complex and wordless awe. Color and the sense of a reality-unlocking secret nearby pervade the experience. There is a sense of other times, and of one's own infancy, and of wonder, wonder, and more wonder. It is an audience with the alien nuncio. In the midst of this experience, apparently at the end of human history, guarding gates that seem surely to open on the howling maelstrom of the unspeakable emptiness between the stars, is the Aeon.

The Aeon, as Heraclitus presciently observed, is a child at play with colored balls. Many diminutive beings are present there—the tykes, the self-transforming machine elves of hyperspace. Are they the children destined to be father to the man? One has the impression of entering into an ecology of souls that lies beyond the portals of what we naively call death. I do not know. Are they the synesthetic embodiment of ourselves as the Other, or of the Other as ourselves? Are they the elves lost to us since the fading of the magic light of childhood? Here is a tremendum barely to be told, an epiphany beyond our wildest dreams. Here is the realm of that which is stranger than we *can* suppose. Here is the mystery, alive, unscathed, still as new for us as when our ancestors lived it fifteen thousand summers ago. The tryptamine entities offer the gift of new language; they sing in pearly voices that rain down as colored petals and flow through the air like hot metal to become toys and such gifts as gods would give their children. The sense of emotional connection is terrifying and intense. The Mysteries revealed are real and if ever fully told will leave no stone upon another in the small world we have gone so ill in.

This is not the mercurial world of the UFO, to be invoked from lonely hilltops; this is not the siren song of lost Atlantis wailing through the trailer courts of crack-crazed America. DMT is not one of our irrational illusions. I believe that what we experience in the presence of DMT is real news. It is a nearby dimension—frightening, transformative, and beyond our powers to imagine, and yet to be explored in the usual way. We must send fearless experts, whatever that may come to mean, to explore and to report on what they find.

DMT, as we have discussed earlier, occurs as a part of ordinary human neurometabolism and is the most powerful of the naturally occurring indole hallucinogens. The extraordinary ease with which DMT utterly destroys all boundaries and conveys one into an impossible-to-anticipate and compellingly Other dimension is one of the miracles of life itself. And this first miracle is followed by a second: the utter ease and simplicity with which enzyme systems in the human brain recognize the DMT molecules at the synapses. After only a few hundred seconds, these enzymes have completely and harmlessly inactivated the DMT and reduced it to by-products of ordinary metabolism. That, with the most powerful of all hallucinogenic indoles, ordinary amine levels in the brain are reestablished so quickly argues there may have been a long co-evolutionary association between human beings and hallucinogenic tryptamines.

Although psilocybin and psilocin, the hallucinogenic indoles active in the cattle-associated *Stropharia cubensis* mushroom, are not presently thought to directly metabolize into DMT before becoming active in the brain, nevertheless their pathway is the closest of relatives to the neural pathway of DMT activity. Indeed, they may be active at the same synapses, with DMT being, however, more reactive. The source of this difference is probably pharmacokinetic—that is, DMT may cross the blood-brain barrier more readily, so that more reaches the site of activity in a shorter time. Affinity of the two compounds for the bond site is approximately equal.

As mentioned earlier, research on DMT, particularly in human beings, has been by and large inadequate. When DMT has been studied, it was administered by injection. This is the preferred procedure with experimental drugs because dosages can be known precisely. Nevertheless, in the case of DMT this approach masked the existence of the extraordinary "turnaround time" of the experience when DMT is smoked. The experience of DMT by intermuscular injection lasts nearly an hour; the peak of the experience obtained by smoking occurs in about one minute. In the Amazon Basin some tribal people have a tradition of using DMT-containing plants. They use the sap of *Virola* trees, relatives of nutmeg, or the ground and toasted seeds of *Anadenanthera peregrina*, a huge leguminous tree. The generally accepted method of activating the indole is to snuff the powdered plant material. Such snuffing is not left to the discretion of the user; rather, the user has a friend blow a hollow reed full of fine powder up first one nostril, then the other (see Figure 27). Excruciating as this process is, it leaves no doubt that Amazonian shamans learned what modern DMT researchers have not: the most effective route of administration is by absorption through the nasal mucosa.

HYPERSPACE AND THE LAW

Perhaps you will object, "But isn't DMT illegal?"

Yes, DMT is currently a Schedule I compound in the United States. Schedule I is a classification for drugs with no proven medical application whatsoever. Not even cocaine rates a Schedule I classification. Psilocybin and DMT were made Schedule I without any scientific evidence at all being presented for or against their use. In the paranoid atmosphere of the late sixties, the mere fact that these compounds cause hallucinations was sufficient grounds for their placement in a category so restrictive that even medical research is discouraged.

Faced with such hysterical Know-Nothingism, we would do well to recall that at one time dissection of corpses was forbidden by the Church and denounced as witchcraft. Modern anatomy was created by medical students who visited battlefields or who stole corpses from the gallows. To advance their knowledge of the human body, they risked arrest and imprisonment. Should we be any less courageous in attempting to push back the frontiers of the known and the possible?

The dominator mentality has always resisted change, almost as if it sensed the possibility of a kind of change that would leave it bereft of its power once and for all. In the phenomenon of the indole hallucinogens that prescient fear has born bounteous fruit—nothing less than the fruit of the Tree of Knowledge. To eat it is to become as gods, and that will surely mean eclipse for the style of the dominators. Such would be the hope of any Archaic Revival.

FIGURE 27. The DMT snuffers. From R. E. Schultes's *Where the Gods Reign* (London: Synergetic Press, 1988), p. 195.

MEETINGS WITH A REMARKABLE OVERMIND

The meltdown of Western rationalism has proceeded quite far, as anyone who will read any up-to-date popular book on cosmology or quantum physics can easily assure himself. Nevertheless, I wish to stoke the fires slightly by adding the concept of some kind of interdimensional nexus that is gained most reliably and directly through the use of indole hallucinogens with a long history of human usage and human coevolution. Such compounds are apparently functioning as regulators of cultural change and can be a means of obtaining access to the intentionality of some very large self-regulating system. Perhaps this is the Overmind of the species, or a kind of "mind of the planet," or perhaps we have been parochial in our search for nonhuman intelligence, and another minded, but radically different, intelligent species shares the earth with us.

I offer these ideas in a speculative vein. I have no strong personal intuition as to what is going on. What I do believe is that I have a sufficient grasp of the customs, expectations, rules of evidence, and "common knowledge" of human beings to be able to report that what is going on inside the DMT intoxication is much more peculiar than anything anyone ever dreamed could be covered by the term "intoxication." When intoxicated by DMT, the mind finds itself in a convincingly real, apparently coexisting alien world. Not a world about our thoughts, our hopes, our fears; rather, a world about the tykes—their joys, their dreams, their poetry. Why? I have not the faintest idea. These are the facts of the matter; this is how it is with us.

Alone among twentieth-century schools of mainstream thought, Jungian psychology has sought to confront some of the phenomena so central to shamanism. Alchemy, which Jung studied very carefully, was the inheritor of a long tradition of shamanistic and magical techniques, as well as more practical chemical procedures such as metalworking and embalming. The literature of alchemy shows that the swirling contents of the alchemical vessel were fertile ground for the projection of the contents of the naive prescientific mind. Jung insisted that alchemical allegories and emblems were products of the unconscious and could be analyzed in the same way as dreams. From Jung's point of view, finding the same motifs in the fantastic speculations of the alchemists and in the dreams of his patients was strong support for his theory of the collective unconscious and its universal generic archetypes.

In the course of his alchemical studies, Jung encountered the accounts of the *cabiri*, the fairylike, alchemical children whose appearance, or felt presence, is a part of the late stages of the alchemical opus.[5] These alchemical children are similar to the small helping spirits that the shaman calls to his aid. Jung saw them as autonomous portions of the psyche that have temporarily escaped from the control of the ego. Unfortunately, the explanation that these alchemical genii are "autonomous portions of the psyche" is no explanation at all. It is as if we were to describe an elf as a small nonphysical person of uncertain parentage. Such explanations only evade the need to confront the deeper nature of experience itself.

Science has not been helpful in the matter of elusive human contacts with other intelligences. It prefers to direct its attention elsewhere, with the comment that subjective experiences, however peculiar, are not its province. What a pity, since subjective experience is all that any of us ever has. Anyhow, the largely subjective nature of the so-called objective universe has now been secured by that most objective of the sciences, physics. The new physics has the subjective observer inextricably tangled with the phenomena observed. Ironically, this is a return to the shamanic point of view. The real intellectual legacy of quantum physics may be the new respectability and primacy that it gives to subjectivity. Recentering ourselves in our subjectivity means a tremendous new reempowering of language, for language is the stuff of which the subjective world is made.

Through psychedelics we are learning that God is not an idea, God is a lost continent in the human mind. That continent has been rediscovered in a time of great peril for ourselves and our world. Is this coincidence, synchronicity, or a cruelly meaningless juxtaposition of hope and ruin? Years ago I directed my life's work toward understanding the mystery at the center of the experience induced by tryptamine hallucinogens. It is not, ultimately, a mystery that science can elucidate. Of course I am aware that one's obsessions expand to fill all space. But in the climactic events surrounding the emergence of pastoralism and language in human beings, I found the ancient echo of the things that I had personally felt and witnessed.

Now the answer sought and found must be faced. Flickering before us is a dimension so huge that its outlines can barely be brought into focus in the human frame of reference. Our animal existence, our planetary existence, is ending. In geological time that end is now only moments away. A great dying, a great extinction of many species, has been occurring since at least the pinnacle of the partnership society in prehistoric Africa. Our future lies in the mind; our weary planet's only hope of survival is that we find ourselves in the mind and make of it a friend that can reunite us with the earth while simultaneously carrying us to the stars. Change, more radical by magnitudes than anything that has gone before, looms immediately ahead. Shamans have kept the gnosis of the accessibility of the Other for millennia; now it is global knowledge. The consequences of this situation have only begun to unfold.

Naturally I do not expect my words to be taken at face value. Nevertheless, these conclusions are based on an experience available to anyone who will but take the time to investigate DMT. The experience itself lasts less than fifteen minutes. I do not anticipate criticism from people who have not taken the trouble to conduct this simple and definitive experiment. After all, how seriously can critics be engaged with the problem if they are unwilling to invest a few minutes of their time to experience the phenomena firsthand?

The deep psychedelic experience does not simply hold out the possibility of a world of sane people living in balance with the earth and one another. It also promises high adventure, engagement with something completely unexpected—a nearby alien universe teeming with life and beauty. Don't ask where; at the present moment we can only say, not here and not there. We have still to admit our ignorance concerning the nature of mind and how precisely the world comes to be and what it is. For more than several millennia our dream has been to understand these matters, and we are defeated. Defeated unless we remember the other possibility—the possibility of the wholly Other.

Some misguided souls scan the heavens for friendly flying saucers that will intervene in profane history and carry us to paradise; others preach redemption at the feet of various rishis, roshis, geysheys, and gurus. Searchers are better advised to look to the work of the botanists, anthropologists, and chemists who have located, identified, and characterized the shamanic hallucinogens. Through them, we have had placed into our hands a tool for the redemption of the human enterprise. It is a great tool, but it is a tool that must be used. Our addictions down through the ages, from sugar to cocaine and television, have been a restless search for the thing torn from us in paradise. The answer has been found. It is no longer something to be sought. It has been found.

RECOVERING OUR ORIGINS

Using plants such as those described above will help us understand the precious gift of plant partnership that was lost at the dawn of time. Many people yearn to be introduced to the facts concerning their true identity. This essential identity is explicitly addressed by a plant hallucinogen. Not to know one's true identity is to be a mad, disensouled thing—a golem. And, indeed, this image, sickeningly Orwellian, applies to the mass of human beings now living in the high-tech industrial democracies. Their authenticity lies in their ability to obey and follow mass style changes that are conveyed through the media. Immersed in junk food, trash media, and cryptofascist politics, they are condemned to toxic lives of low awareness. Sedated by the prescripted daily television fix, they are a living dead, lost to all but the act of consuming.

I believe that the failure of our civilization to come to terms with the issue of drugs and habitual destructive behavior is a legacy of unhappiness for us all. But if we sufficiently reconstructed our image of self and world, we could make out of psychopharmacology the stuff of our grandest hopes and dreams. Instead, pharmacology has become the demonic handmaiden of an unchecked descent into regimentation and erosion of civil liberties.

Most people are addicted to some substance and, more important, all people are addicted to patterns of behavior. Attempting to distinguish between habits and addictions does damage to the indissoluble confluence of mental and physical energies that shapes the behavior of each of us. People not involved in a relationship with food/drug stimulation are rare and by their preference for dogma and deliberately self-limited horizons must be judged to have failed to create a viable alternative to substance involvement.

I have attempted here to examine our biological history and our more recent cultural history with an eye to something that may have been missed. My theme was human arrangements with plants, made and broken over the millennia. These relationships have shaped every aspect of our identities as self-reflecting beings—our languages, our cultural values, our sexual behavior, what we remember and what we forget about our own past. Plants are the missing link in the search to understand the human mind and its place in nature.

In the United States, the federal government's zeal to appear to wish to eradicate drugs is directly linked to the degree to which the government has been co-opted by the values of fundamentalist Christianity. We entertain the illusion of the constitutional separation of church and state in the United States. But in fact the federal government, when it acted to prohibit alcohol during Prohibition, when it interferes with rights to reproductive freedom, or with the use of peyote in Native American religious rituals, and when it attempts unreasonably to regulate foods and substances, is acting as the enforcing arm for the values of right-wing fundamentalism.

Eventually the right to determine our own food and drug preferences will be seen as a natural consequence of human dignity, as long as it is done in a way that does not limit the rights of others. The signing of the Magna Carta, the abolition of slavery, the enfranchisement of women—these are instances in which the evolving definition of what constituted fairness swept away ossified social structures that had come to rely more and more on a "fundamentalist" reading of their own first principles. The war on drugs is schizophrenically waged by governments that deplore the drug trade and yet are also the major guarantors and patrons of the international drug cartels. Such an approach is doomed to failure.

The war on drugs was never meant to be won. Instead, it will be prolonged as long as possible in order to allow various intelligence operations to wring the last few hundreds of millions of dollars in illicit profits from the global drug scam; then defeat will have to be declared. "Defeat" will mean, as it did in the case of the Vietnam War, that the media will correctly portray the true dimensions of the situation and the real players, and that public revulsion at the culpability, stupidity, and venality of the Establishment's role will force a policy review. In cynically manipulating nations and peoples with narcotics and stimulants, modern governments have associated themselves with an ethical disaster comparable to the eighteenth-century rebirth of the slave trade or the recently renounced excesses of Marxism-Leninism.

The conclusion seems obvious: only legalization can lay the basis for a sane drug policy. Indeed, this opinion has been reached by most disinterested commentators on the problem, although the political consequences of advocating legalization have made it slow to be considered. Most recently Arnold Trebach's thoughtful book, *The Great Drug War*, has marshaled persuasive arguments in favor of a revolution in drug policy:

> Another model for guidance in approaching the subject of drug abuse may be found in the manner with which America has historically dealt with conflicting religious creeds; virtually all are accepted as decent moral options that ought to be available for those people who believe in them. The subject of drugs should be approached in the same spirit—more like religion than science. My wish is that law and medicine recognize the personal and nonscientific nature of the drug-abuse arena by enacting some form of First Amendment guarantee of freedom to select a personal drug abuse doctrine, but limited somewhat by enlightened principles of medicine.[6]

What Trebach does not discuss, indeed does not even mention, is the role to be played by hallucinogens in the postsuppression scenario. Indeed, psychedelics seem unimportant if the only measure of a drug's social impact is the estimate of the millions of dollars of street sales that may have taken place. Only LSD continues to be occasionally singled out among the psychedelics as a possible large-scale problem. However, estimates of the amount of psychedelics produced and used in the United States have been politicized and hence are unreliable and meaningless.

But another measure of the social importance of a substance argues that we are remiss in not at least beginning to discuss the social impact of psychedelic use when we contemplate legalization of drugs. A clue to that other measure is the interest the CIA and military intelligence gave to psychedelics during the sixties through projects such as MK (for mind control) and MK-ULTRA. The widespread belief that the conclusion of these studies was that television was the drug of choice for mass hypnosis, while reasonable, should not be taken at face value. I believe that, once drugs are legalized, the fear that there will be a vast epidemic of cocaine or heroin addiction will be proven groundless. I also believe that there will be increased interest in and use of psychedelics, and that this possibility is of great concern to the Establishment. This new interest in psychedelics should be anticipated and provided for. If use of psychedelics makes it easier to recapture the social attitudes and assumptions of the original partnership cultures, then eventually educational institutions may wish to encourage this awareness.

A new global consensus appears to be building. What was previously inchoate and unconscious is now becoming conscious and at the same time structured. The collapse of the Marxist alternative to media-dense, high-tech democratic consumerism has been swift and complete. For the first time in planetary history, a defined, albeit dimly defined, consensus exists for "democratic values." This trend will encounter real resistance from various forms of monotheistic religious fundamentalism during the 1990s. It is a phenomenon of expanded consciousness driven by the information explosion. Democracy is an articulation of the Archaic notion of the nomadic egalitarian group. In its purest expression it is thoroughly psychedelic and its triumph seems ultimately certain.

The "drug problem" runs against the tendency toward global expansion of consciousness through spread of democratic values. There is no question that a society that sets out to control its citizens' use of drugs sets out on the slippery path to totalitarianism. No amount of police power, surveillance, and intrusion into people's lives can be expected to affect "the drug problem." Hence there is no limit to the amount of repression that frightened institutions and their brainwashed populations may call for.

A MODEST PROPOSAL

A drug policy respectful of democratic values would aim to educate people to make informed choices based on their own needs and ideals. Such a simple prescription is necessary and sadly overdue.

A master plan for seriously seeking to come to terms with America's drug problems might explore a number of options, including the following.

1. A 200 percent federal tax should be imposed on tobacco and alcohol. All government subsidies for tobacco production should be ended. Warnings on packaging should be strengthened. A 20 percent federal sales tax should be levied on sugar and sugar substitutes, and all supports for sugar production should be ended. Sugar packages should also carry warnings, and sugar should be a mandatory topic in school nutrition curricula.
2. All forms of cannabis should be legalized and a 200 percent federal sales tax imposed on cannabis products. Information as to the THC content of the product and current conclusions regarding its impact on health should be printed on the packaging.
3. International Monetary Fund and World Bank lending should be withdrawn from countries that produce hard drugs. Only international inspection and certification that a country is in compliance would restore loan eligibility.
4. Strict gun control must apply to both manufacture and possession. It is the unrestricted availability of firearms that has made violent crime and the drug abuse problem so intertwined.
5. The legality of nature must be recognized, so that all plants are legal to grow and possess.
6. Psychedelic therapy should be made legal and insurance coverage extended to include it.
7. Currency and banking regulations need to be strengthened. Presently bank collusion with criminal cartels allows large-scale money laundering to take place.
8. There is an immediate need for massive support for scientific research into all aspects of substance use and abuse and an equally massive commitment to public education.
9. One year after implementation of the above, all drugs still illegal in the United States should be decrimi-

nalized. The middleman is eliminated, the government can sell drugs at cost plus 200 percent, and those monies can be placed in a special fund to pay the social, medical, and educational costs of the legalization program. Money from taxes on alcohol, tobacco, sugar, and cannabis can also be placed in this fund.

10. Also following this one-year period, pardons should be given to all offenders in drug cases that did not involve firearms or felonious assault.

If these proposals seem radical, it is only because we have drifted so far from the ideals that were originally most American. At the foundation of the American theory of social polity is the notion that our inalienable rights include "life, liberty, and the pursuit of happiness." To pretend that the right to the pursuit of happiness does not include the right to experiment with psychoactive plants and substances is to make an argument that is at best narrow and at worst ignorant and primitive. The only religions that are anything more than the traditionally sanctioned moral codes are religions of trance, dance ecstasy, and intoxication by hallucinogens. The living fact of the mystery of being is there, and it is an inalienable religious right to be able to approach it on one's own terms. A civilized society would enshrine that principle in law.

EPILOGUE: LOOKING OUTWARD AND INWARD TO A SEA OF STARS

We have arrived at the point in our story where history merges with the political energies of the moment. The current controversies that have use and abuse of substances as their theme must share the stage with other issues of equal import: poverty and overpopulation, environmental destruction, and unmet political expectations. These phenomena are the inevitable by-products of the dominator culture. In struggling with these social problems we must remember that the roots of our humanness lie elsewhere, in the cascade of mental abilities that were unleashed within our species many tens of millennia ago—the ability to name, to classify, to compare, and to remember. These functions all can be traced back to the quasi-symbiotic relationship that we enjoyed with psilocybin mushrooms in the African partnership society of prehistory.

Our breach of faith with the symbiotic relationship to the plant hallucinogens has made us susceptible to an ever more neurotic response to each other and the world around us. Several thousand years of such bereavement have left us the nearly psychotic inheritors of a planet festering with the toxic by-products of scientific industrialism.

IF NOT US, WHO? IF NOT NOW, WHEN?

It is time for us to undertake a dialogue based on an objective assessment of what our culture does and means. Another hundred years of business as usual is inconceivable. Dogma and ideology have become obsolete; their poisonous assumptions allow us to close our eyes to our hideous destructiveness and to loot even those resources that properly belong to our children and grandchildren. Our toys do not satisfy; our religions are no more than manias; our political systems are a grotesque aping of what we intended them to be.

How can we hope to do better? Although fears of nuclear confrontation have diminished with the recent changes in the Eastern bloc, the world is still plagued by hunger, overpopulation, racism, sexism, and religious and political fundamentalism. We have the capacity—industrial, scientific, and financial—to change the world. The question is, do we have the capacity to change ourselves, to change our minds? I believe that the answer to this must be yes but not without help from nature. If mere preaching of virtue could provide the answer, then we would have arrived at the threshold of angelic existence some time ago. If mere legislation of virtue were an answer, we would have learned that a long time ago.

Help from nature means recognizing that the satisfaction of the religious impulse comes not from ritual, and still less from dogma, but rather, from a fundamental kind of experience—the experience of symbiosis with hallucinogenic plants and, through them, symbiosis with the whole of planetary life. Radical as this proposal may appear, it has been anticipated in the work of no less a sober observer of Western culture than Arthur Koestler:

> Nature has let us down, God seems to have left the receiver off the hook, and time is running out. To hope for salvation to be synthesized in the laboratory may seem materialistic, crankish, or naive; but, to tell the truth, there is a Jungian twist to it—for it reflects the ancient alchemist's dream to concoct the *elixir vitae*. What we expect from it, however, is not eternal life, nor the transformation of base

> metal into gold, but the transformation of *homo maniacus* into *homo sapiens*. When man decides to take his fate into his own hands, that possibility will be within reach.[1]

Koestler concludes from his examination of our history of institutionalized violence as a species that some form of pharmacological intervention will be necessary before we can be at peace with one another. He proceeds to make an argument for conscious and scientifically managed psychopharmacological intervention in the life of society that has grave implications for the preservation of ideals of human independence and liberty. Koestler was apparently unaware of the shamanic tradition or of the richness of the psychedelic experience. Hence he was not aware that the task of managing a global human population into a state of balance and happiness could involve introducing the experience of an internal horizon of transcendence into people's lives.

FINDING THE WAY OUT

Without the escape hatch into the transcendental and transpersonal realm that is provided by plant-based indole hallucinogens, the human future would be bleak indeed. We have lost the ability to be swayed by the power of myths, and our history should convince us of the fallacy of dogma. What we require is a new dimension of self-experience that individually and collectively authenticates democratic social forms and our stewardship of this small part of the larger universe.

Discovery of such a dimension will mean risk and opportunity. Seeking the answer is the stance of the ingenue, the preinitiate, and the fool. We must now have done with such posturing; it is for us to face the answer. Facing the answer means recognizing that the world we have prepared to hand on to the generations of the future is no more than a mess of broken pottage. It is not the dispossessed people of the ruined rain forests who are pathetic, it is not the stoic opium farmers of tribal Burma who menace distant hopes and populations—it is ourselves.

FROM THE GRASSLANDS TO THE STARSHIP

Human history has been a fifteen-thousand-year dash from the equilibrium of the African cradle to the twentieth-century apotheosis of delusion, devaluation, and mass death. Now we stand on the brink of star flight, virtual reality technologies, and a revivified shamanism that heralds the abandonment of the monkey body and tribal group that has always been our context. The age of the imagination is dawning. The shamanic plants and the worlds that they reveal are the worlds from which we imagine that we came long ago, worlds of light and power and beauty that in some form or another lie behind the eschatological visions of all of the world's great religions. We can claim this prodigal legacy only as quickly as we can remake our language and ourselves.

Remaking our language means rejecting the image of ourselves inherited from dominator culture—that of a creature guilty of sin and hence deserving of exclusion from paradise. Paradise is our birthright and can be claimed by any one of us. Nature is not our enemy, to be raped and conquered. Nature is ourselves, to be cherished and explored. Shamanism has always known this, and shamanism has always, in its most authentic expressions, taught that the path required allies. These allies are the hallucinogenic plants and the mysterious teaching entities, luminous and transcendental, that reside in that nearby dimension of ecstatic beauty and understanding that we have denied until it is now nearly too late.

WE AWAIT OURSELVES WITHIN THE VISION

We can now move toward a new vision of ourselves and our role in nature. We are the omni-adaptable species, we are the thinkers, the makers, and the solvers of problems. These great gifts that are ours alone and which come out of the evolutionary matrix of the planet are not for us—our convenience, our satisfaction, our greater glory. They are for life; they are the special qualities that we can contribute to the great community of organic being, if we are to become the care giver, the gardener, and the mother of our mother, which is the living earth.

Here there is great mystery. In the middle of the slow-moving desert of unreflecting nature we come upon ourselves and perhaps see ourselves for the first time. We are colorful, cantankerous, and alive with hopes and dreams that, so far as we know, are unique in the universe. We have been too long asleep and shackled by the power we have ceded to the least noble parts of ourselves and the least noble among us. It is time that we stood up and faced the fact that we must and *can* change our minds.

The long night of human history is drawing at last to its conclusion. Now the air is hushed and the east is streaked with the rosy blush of dawn. Yet in the world we have always known evening grows deeper and the shadows lengthen toward a night that will know no end. One way or another the story of the foolish monkey is nearly forever over. Our destiny is to turn without regret from what has been, to face ourselves, our parents, lovers, and children, to gather our tool kits, our animals, and the old, old dreams, so that we may move out across the visionary landscape of ever-deeper understanding. Hopefully there, where we have always been most comfortable, most ourselves, we will find glory and triumph in the search for meaning in the endless life of the imagination, at play at last in the fields of an Eden refound.

NOTES

INTRODUCTION

1. See Alfred W. McCoy, *The Politics of Heroin in Southeast Asia*, (New York: Harper Colophon Books, 1972), who observes on p. 16:

With American consumer demand [for heroin] reduced to its lowest point in fifty years and the international syndicates in disarray, the U.S. government had a unique opportunity to eliminate heroin addiction as a major American social problem. However, instead of delivering the death blow to these criminal syndicates, the U.S. government—through the Central Intelligence Agency and its wartime predecessor, the OSS—created a situation that made it possible for the Sicilian-American Mafia and the Corsican underworld to revive the international narcotics traffic.

2. Victor Marchetti and John D. Marks, *The CIA and the Cult of Intelligence* (New York: Knopf, 1974), p. 256. See also H. Kruger (1980) and A. W. McCoy (1972).

3. Ronald K. Siegel, *Intoxication* (New York: E. P. Dutton, 1989), p. 119.

4. Riane Eisler, *The Chalice and the Blade* (San Francisco: Harper & Row, 1987).

1. SHAMANISM: SETTING THE STAGE

1. See Mircea Eliade, *Shamanism: Archaic Techniques of Ecstasy* (New York: Pantheon, 1964), pp. 23 ff.

2. Dennis McKenna and Terence McKenna, *The Invisible Landscape* (New York: Seabury Press, 1975), p. 10.

3. Eliade (1959), p. 9.

4. Quoted in Roger Lewin, *In the Age of Mankind* (New York: Smithsonian Institution, 1988), p. 180.

5. Cf. William Burroughs and A. Ginsberg, *The Yagé Letters* (San Francisco: City Lights Books, 1963).

2. THE MAGIC IN FOOD

1. E. Rodriguez, M. Aregullin, S. Uehara, T. Nishida, R. Wrangham, Z. Abramowski, A. Finlayson, and G. H. N. Towers, "Thiarubrine-A, A Bioactive Constituent of *Aspilia (Asteraceae)* Consumed by Wild Chimpanzees," *Experientia* 41 (1985): 419–420.

2. Edward O. Wilson, *Biophilia* (Cambridge, Mass.: Harvard University Press, 1984), p. 33.

3. Charles J. Lumsden and Edward O. Wilson, *Promethean Fire: Reflections on the Origin of Mind* (Cambridge, Mass.: Harvard University Press, 1983), p. 12.

4. Ibid., p. 15.

5. Roland Fischer, et al., "Psilocybin-Induced Contraction of Nearby Visual Space," *Agents and Actions* 1, no. 4 (1970): 190–197.

6. Dennis McKenna, "Hallucinogens and Evolution." Seminar transcript abstract, given in 1984, Esalen, p. 2.

3. THE SEARCH FOR THE ORIGINAL TREE OF KNOWLEDGE

1. A. Hoffer and H. Osmond, *New Hope for Alcoholics* (New York: University Books, 1968).

2. Carl Saur, *Man's Impact on the Earth* (New York: Academic Press, 1973).

3. James W. Fernandez, *Bwiti: An Ethnography of the Religious Imagination in Africa* (Princeton: Princeton University Press, 1982).

4. Gracie and Zarkov, "An Indo-European Plant Teacher," *Notes from Underground* 10 (Berkeley).

5. O. T. Oss and O. N. Oeric, *Psilocybin: The Magic Mushroom Grower's Guide* (Berkeley: Lux Natura Press, 1986).

4. PLANTS AND PRIMATES: POSTCARDS FROM THE STONED AGE

1. Herbert V. Guenther, *Tibetan Buddhism without Mystification* (Leiden, Netherlands: E. J. Brill, 1986), p. 66.

2. Francisco J. Varela and A. Coutinho "The Body Thinks: How and Why the Immune System Is Cognitive" in *The Reality Club*, ed. John Brockman, vol. 2. (New York: Phoenix Press, 1988).

3. Aldous Huxley, *The Doors of Perception* (New York: Harper, 1954), p. 22.

4. Julian Jaynes, *The Origin of Consciousness in the Breakdown of the Bicameral Mind* (Boston: Houghton Mifflin, 1977).

5. Henry Munn, "The Mushrooms of Language," in Michael J. Harner, ed., *Shamanism and Hallucinogens* (London: Oxford University Press, 1973), p. 88.

6. K. F. Jindrak and H. Jindrak, "Mechanical Effect of Vocalization of Human Brain and Meninges," *Medical Hypotheses* 25 (1988), pp. 17–20.

5. HABIT AS CULTURE AND RELIGION

1. R. Gordon Wasson, Albert Hofmann, and Carl Ruck, *The Road to Eleusis* (New York: Harcourt Brace Jovanovich, 1978), p. 23.

2. C. H. Waddington, *The Nature of Life* (London: Allen & Unwin, 1961).

3. Mircea Eliade, *Yoga: Immortality and Freedom* (New York: Pantheon, 1958), p. 320.

4. R. Gordon Wasson, *The Wondrous Mushroom: Mycolatry in Mesoamerica* (New York: McGraw-Hill, 1980), p. 225.

5. For further rebuttal of Eliade's position see also R. Gordon Wasson, *Soma: Divine Mushroom of Immortality* (New York: Harcourt Brace Jovanovich, 1971), pp. 326–334.

6. James W. Fernandez, *Bwiti: An Ethnography of the Religious Imagination in Africa* (Princeton: Princeton University Press, 1982), p. 311.

7. Christian Rätsch and Claudia Müller-Ebeling, *Isoldens Liebestrank Aphrodisiaka in Geschichte und Gegenwart* (Munich: Kindler Verlag, 1986).

8. Jean Baker Miller, *Toward a New Psychology of Women* (Boston: Beacon Press, 1986).

6. THE HIGH PLAINS OF EDEN

1. This connection between the Tassili art and mushroom use was pointed out to me by Jeff Gaines, an ethnomycologist and art historian living in Boulder, Colorado. He recognized the implications of the Tassili images for the role of mushroom use in human prehistory. As we go to press it has been called to my attention that I have recently been joined in my opinions concerning the existence of Archaic mushroom use by the Roundhead Culture of the Tassili Plateau. An Italian scholar, Giorgio Samorini, has called attention to the presence of mushroom motifs in the Tassili rock art and assumes a mushroom cult of great antiquity to have flourished in the area. Cf. G. Samorini, 1989, pp. 18–22, see bibliography. See also Roger Lewin, "Stone Age Psychedelia" in *New Scientist*, 8 June 1991, pp. 30–34.

2. Lionel Balout, *Algérie Préhistorique* (Paris: Arts et Métiers Graphiques, 1958).

3. John E. Pfeiffer, *The Creative Explosion: An Inquiry into the Origins of Art and Religion* (Ithaca, N.Y.: Cornell University Press, 1982), p. 213.

4. James Mellaart, *Earliest Civilizations of the Near East* (New York: McGraw-Hill, 1965), p. 29.

5. Mary Settegast, *Plato Prehistorian* (Cambridge: Rotenberg Press, 1987), p. 154.

6. Donald Owen Henry, *The Natufian of Palestine: Its Material Culture and Ecology* (Ann Arbor: University Microfilms, 1973), p. v.

7. D. A. E. Garrod, "The Natufian Culture: The Life and Economy of a Mesolithic People in the Near East," *Proceedings of the British Academy* 43 (1957): 211–227.

8. Settegast, op. cit., p. 2.

9. Mellaart, op. cit.; Mellaart, *Çatal Hüyük: A Neolithic Town in Anatolia* (New York: McGraw-Hill, 1967), pp. 221–222

10. Ibid., p. 226.

11. Settegast, op. cit., pp. 166–167.

12. Mellaart, *Earliest Civilizations*, p. 79.

13. Settegast, op. cit., p. 180.

14. Mellaart, *Earliest Civilizations*, p. 77.

15. Marija Gimbutas, *The Goddesses and Gods of Old Europe* (Berkeley: University of California Press, 1982).

16. Colin Renfrew, *Archaeology and Language: The Puzzle of Indo-European Origins* (London: Cambridge University Press, 1988), p. 171.

17. Vitaly Shevoroshkin, "The Mother Tongue," *The Sciences*, May/June 1990, pp. 20–27.

18. R. Gordon Wasson, Albert Hofmann, and Carl Ruck, *The Road to Eleusis* (New York: Harcourt Brace Jovanovich, 1978).

19. Hans Peter Duerr, *Dreamtime: Concerning the Boundary between Wilderness and Civilization* (Oxford: Basil Blackwell, 1985), p. 4.

7. SEARCHING FOR SOMA: THE GOLDEN VEDIC ENIGMA

1. At Anuvaka V, X, 5, 1.

2. H. H. Wilson, trans., *Rig-Veda Sanhita* (Poona, India: Ashtekar, 1928), vol. 5, p. 287.

3. Adolf Kaegi, *The Rig Veda: The Oldest Literature of the Indians* (Boston: Ginn, 1886), pp. 72–73.

4. R. C. Zahner, *The Dawn and Twilight of Zoroastrianism* (New York: Putnam, 1961), p. 86.

5. H. D. Griswold, *The Religion of the Rigveda* (London: Oxford University Press, 1923).

6. Zahner, op. cit., p. 86.
7. George W. Cox, *The Mythology of the Aryan Nations* (London: C. Kegan Paul, 1878), pp. 387–388.
8. See Ernest Bender, ed., *R. Gordon Wasson on Soma and Daniel H. H. Ingalls' Response* (Publication #7), (New Haven: American Oriental Society, 1971).
9. David Flattery and Martin Schwartz, *Haoma and Harmaline*, Near Eastern Studies, vol. 21 (Berkeley: University of California Press, 1989).
10. Ibid., section 31.
11. R. Gordon Wasson, *Soma: Divine Mushroom of Immortality* (New York: Harcourt Brace Jovanovich, 1971), p. 75.
12. Martha Windholz, ed., *The Merck Index*, 9th ed. (Rahway, N.J.: Merck, 1976).
13. Personal communication, 1988.
14. R. Gordon Wasson, *Persephone's Quest: Entheogens and the Origins of Religion* (New Haven: Yale University Press, 1986), p. 135.
15. Ibid., p. 134.
16. R. Gordon Wasson, *The Wondrous Mushroom: Mycolatry in Mesoamerica* (New York: McGraw-Hill, 1980), p. xvi.
17. Flattery and Schwartz, op. cit.
18. S. H. Hooke, *Babylonian and Assyrian Religion* (Norman: University of Oklahoma Press, 1963), p. 19.
19. Gaston Maspero, *The Dawn of Civilization—Egypt and Chaldea* (London: Society for Promoting Christian Knowledge, 1894), p. 655.
20. Wasson, *Soma*, p. 256.
21. Carl Ruck, co-author, *The Road to Eleusis* (New York: Harcourt Brace Jovanovich, 1978), in Wasson, *Persephone's Quest*, p. 256.
22. R. Gordon Wasson, personal communication, 1977.
23. Cf. Marija Gimbutas, *The Goddesses and Gods of Old Europe, 6500–3500 B.C.: Myths and Cult Images* (Berkeley: University of California Press, 1982), p. 219.

8. TWILIGHT IN EDEN: MINOAN CRETE AND THE ELEUSINIAN MYSTERY

1. Riane Eisler, *The Chalice and the Blade: Our History, Our Future* (San Francisco: Harper & Row, 1987), pp. 30–31.
2. Martin P. Nilsson, *A History of Greek Religion* (New York: W. W. Norton, 1964), p. 13.
3. Carl Kerényi, *Dionysos: Archetypal Image of Indestructible Life*, Bollingen Series LXV (Princeton: Princeton University Press, 1976), p. 17.
4. Cf. Elmer G. Suhr, *Before Olympos* (New York: Helios Books, 1967).
5. In Lykophron, 811; Tzetes, Scholia on Lykophron, 798; and Eustathios, Scholia on Homer, 369, 20, and 894, 42.

6. Bibliotheke III, 3 f.

7. Axel W. Persson, *The Religion of Greece in Prehistoric Times* (Berkeley: University of California Press, 1942), p. 10.

8. R. Gordon Wasson, *Soma: Divine Mushroom of Immortality* (New York: Harcourt Brace Jovanovich, 1971), p. 16.

9. Kerényi, op. cit., p. 27.

10. Walter F. Otto, *Dionysus Myth and Cult* (Bloomington: Indiana University Press, 1965), p. 65.

11. Robert Graves, *The Greek Myths*, 2 vols. (Baltimore: Penguin, 1955), p. 57.

12. Persson, op. cit., p. 150.

13. Le Clerc de Septchenes, *The Religion of the Ancient Greeks* (London: Elliot and T. Kay, 1788), p. 180.

14. Walter F. Otto, "The Meaning of the Eleusinian Mysteries," in Joseph Campbell, *Eranos Yearbook Number Two: The Mysteries* (New York: Pantheon, 1955), p. 23.

15. Robert Graves, *Difficult Questions, Easy Answers* (Garden City, N.Y.: Doubleday, 1964), pp. 106–107.

16. R. Gordon Wasson, Albert Hofmann, and Carl Ruck, *The Road to Eleusis* (New York: Harcourt Brace Jovanovich, 1978).

17. Kerényi, op. cit., p. 53.

18. Mary Allerton Kilbourne Matossian, *Poisons of the Past: Molds, Epidemics, and History* (New Haven: Yale University Press, 1989).

19. A. Hoffer and H. Osmond, *The Hallucinogens* (New York: Academic Press, 1967), p. 84.

9. ALCOHOL AND THE ALCHEMY OF SPIRIT

1. Aldous Huxley, *Moksha: Writings on Psychedelics and the Visionary Experience*, Michael Horowitz and Cynthia Palmer, eds. (Los Angeles: Tarcher, 1977), p. 97.

2. Fasti III 736.

3. Carl Kerényi, *Dionysos: Archetypal Image of Indestructible Life*, Bollingen Series LXV (Princeton: Princeton University Press, 1976), p. 98.

4. E. S. Drower, *Water into Wine* (London: John Murray, 1956), p. 7.

5. Ronald K. Siegel, *Intoxication* (New York: E. P. Dutton, 1989), p. 119.

6. James R. Ware, *Alchemy, Medicine, Religion in the China of A.D. 320: The Nei Pien of Ko Hung* (Cambridge, Mass.: MIT Press), p. 72.

7. Richard R. Matheson, *The Eternal Search: The Story of Man and His Drugs* (New York: G. P. Putnam's Sons, 1958).

8. Charles H. La Wall, *The Curious Lore of Drugs and Medicines through the Ages* (Philadelphia: J. B. Lippincott, 1927), p. 158.

9. Lewis Lewin, *Phantastica: Narcotic and Stimulating Drugs* (New York: E. P. Dutton, 1931), p. 190.

10. THE BALLAD OF THE DREAMING WEAVERS: CANNABIS AND CULTURE

1. See Robert P. Walton, *Marijuana: America's New Drug Problem* (Philadelphia: J. B. Lippincott, 1938), pp. 188–195.

2. Herodotus, *Works*, H. Cary, trans. (London: George Bell and Sons, 1901), Book IV, Chapter 74.

3. Herodotus, op. cit., Book I, Chapter 202.

4. Julian quoted in Walton, op. cit., p. 3.

5. J. Campbell Oman, *The Mystics, Ascetics, and Saints of India* (London: T. Fischer Unwin, 1903).

6. Quoted in Walton, op. cit., p. 8.

7. J. F. de Lacroix, *Anecdotes Arabes et Musulmanes, Depuis l'An de J.C. 614* (Paris: Vincent, 1772), p. 534.

8. Bayard Taylor, *The Lands of the Saracen* (New York: G. P. Putnam, 1855), pp. 137–139.

9. Fitz Hugh Ludlow, *The Hasheesh Eater: Being Passages from the Life of a Pythagorean* (New York: Harper & Brothers, 1857), p. 86.

10. Ibid., pp. 288–289.

11. Herer, Jack, *The Emperor Wears No Clothes* (Business Alliance for Commerce in Hemp [BACH], 1990).

11. COMPLACENCIES OF THE PEIGNOIR: SUGAR, COFFEE, TEA, AND CHOCOLATE

1. De Lacy O'Leary, *How Greek Science Passed to the Arabs* (London: Routledge & Kegan Paul, 1949), p. 71.

2. Henry Hobhouse, *Seeds of Change: Five Plants That Transformed Mankind* (New York: Harper & Row, 1985), p. 46.

3. Janice Keller Phelps and Alan E. Nourse, *The Hidden Addiction and How to Get Free* (Boston: Little, Brown, 1986), p. 75.

4. Hobhouse, op. cit., p. 54.

5. Hobhouse, op. cit., p. 63.

6. Wallace Stevens, *The Collected Poems of Wallace Stevens* (New York: Alfred A. Knopf, 1981).

7. Hobhouse, op. cit., pp. 96–97.

8. Hobhouse, op. cit., p. 108.

9. Lewis Lewin, *Phantastica: Narcotic and Stimulating Drugs* (New York: E. P. Dutton, 1931), pp. 256–257.

10. Ibid., pp. 257–258.

11. Jonathan Ott, *The Cacahuatl Eater: Ruminations of an Unabashed Chocolate Eater* (Vashon, Washington: Natural Products Co., 1985), pp. 12–22.

12. O. T. Oss and O. N. Oeric, *Psilocybin: The Magic Mushroom Grower's Guide* (Berkeley: Lux Natura Press, 1986), p. 73.

13. Lewin, op. cit., p. 283.

12. SMOKE GETS IN YOUR EYES: OPIUM AND TOBACCO

1. Arnold S. Trebach, *The Great Drug War* (New York: Macmillan, 1987), p. 291.
2. Carl Kerényi, *Dionysos: Archetypal Image of Indestructible Life*, Bollingen Series LXV (Princeton: Princeton University Press, 1976), p. 23.
3. William Emboden, *Narcotic Plants* (New York: Macmillan, 1979), pp. 27–28.
4. Alethea Hayter, *Opium and the Romantic Imagination* (Berkeley: University of California Press, 1968), p. 22.
5. Fred Gettings's *Dictionary of Occult, Hermetic, and Alchemical Sigils* (London: Routledge & Kegan Paul, 1981) contains no sigil or special marking for opium though it contains such marks for hundreds of other substances and materials.
6. Quoted in Lewin, p. 38.
7. Lewis Lewin, *Phantastica: Narcotic and Stimulating Drugs* (New York: E. P. Dutton, 1931), p. 288.
8. Francis Robicsek, *The Smoking Gods: Tobacco in Maya Art, History and Religion* (Norman: University of Oklahoma Press, 1978), p. 46.
9. Peter Furst, *Hallucinogens and Culture* (San Francisco: Chandler & Sharp, 1978), p. 28.
10. Thomas Bartholin, *Historiarum anatomicarum et medicarum rariorum* (Copenhagen, 1661).
11. Emboden, op. cit., p. 38.
12. Robiscek, op. cit., p. 8.
13. Henry Hobhouse, *Seeds of Change: Five Plants That Transformed Mankind* (New York: Harper & Row, 1985), p. 117.
14. Arthur Waley, *The Opium War Through Chinese Eyes* (Stanford: Stanford University Press, 1958), pp. 11–157.
15. Peter Ward Fay, *The Opium War* (New York: W.W. Norton, 1975), pp. 249–260. Also see Jack Beeching, *The Chinese Opium Wars* (New York: Harcourt Brace Jovanovich, 1975).
16. Thomas De Quincey, *Confessions of an English Opium-Eater* (London: MacDonald, 1822), p. 117.
17. Hayter, op. cit., p. 103.

13. SYNTHETICS: HEROIN, COCAINE, AND TELEVISION

1. William Burroughs, *Naked Lunch* (New York: Grove Press, 1959), p. viii.
2. James A. Duke, David Aulik, and Timothy Plowman, "Nutritional Value of Coca," *Botanical Museum Leaflets of Harvard University* 24:6 (1975).
3. Sigmund Freud, *The Cocaine Papers* (Vienna: Dunquin Press, 1963), p. 14.
4. Arthur Conan Doyle, *The Sign of Four* in *The Complete Sherlock Holmes* (New York: Doubleday, 1905).

5. Introduction by Michael Horowitz, in W. Golden Mortimer, *History of Coca, the Divine Plant of the Incas* (San Francisco: Fitz Hugh Ludlow Library, 1974).

6. Alfred W. McCoy, *The Politics of Heroin in Southeast Asia* (New York: Harper Colophon Books, 1972).

7. Henrik Krüger, *The Great Heroin Coup: Drugs, Intelligence and International Fascism* (Boston: South End Press, 1980), p. 14

8. Philip K. Dick, *The Man in the High Castle* (London: Penguin, 1965).

9. Marie Winn, *The Plug-In Drug* (New York: Penguin, 1977), pp. 24–25.

10. Jerry Mander, *Four Arguments for the Elimination of Television* (New York: Quill, 1978), p. 197.

11. Martin A. Lee and Bruce Shlain, *Acid Dreams: The CIA, LSD, and the Sixties Rebellion* (New York: Grove Press, 1985), pp. 27–35.

14. A BRIEF HISTORY OF PSYCHEDELICS

1. Hans Peter Duerr, *Dreamtime: Concerning the Boundary Between Wilderness and Civilization* (Oxford: Basil Blackwell, 1985).

2. Richard Spruce, *Notes of a Botanist on the Amazon and Rio Negro*, A. R. Wallace, ed. (London: Macmillan, 1980).

3. Richard Evans Schultes, "The Beta-Carboline Hallucinogens of South America," *Journal of Psychoactive Drugs* 14, no. 3 (1982): 205–220.

4. Claudio Naranjo, *The Healing Journey: New Approaches to Consciousness* (New York: Ballantine, 1973); Marlene Dobkin de Rios, *Visionary Vine: Psychedelic Healing in the Peruvian Amazon* (San Francisco: Chandler, 1972); Luis Eduardo Luna, *Vegetalismo: Shamanism among the Mestizo Population of the Peruvian Amazon* (Stockholm: Alquist & Wiksell, 1986).

5. Quoted in A. Hoffer and H. Osmond, *The Hallucinogens* (New York: Academic Press, 1967), p. 8.

6. Ibid., p. 9.

7. Ibid., p. 7.

8. Heinrich Klüver, *Mescal, the Divine Plant and Its Psychological Effects* (London: Kegan Paul, 1928), p. 28.

9. Cf. Victor A. Reko, *Magische Gife, Rausch-und Betäubungsmittel der neuen Welt* (Berlin: Express Edition, 1987).

10. Richard Evans Schultes, "Plantae Mexicanae, II: The Identification of Teonanácatl, a Narcotic Basidiomycete of the Aztecs," *Botanical Museum Leaflets of Harvard University* (1939) 7:37–54.

11. Alfred Jarry, *Selected Works of Alfred Jarry*, Roger Shattuck and Simon Watson Taylor, eds. (New York: Grove Press, 1965).

12. Albert Hofmann, *LSD My Problem Child* (Los Angeles: Tarcher, 1983), p. 15.

13. Aldous Huxley, *The Doors of Perception* (New York: Harper, 1954), p. 33.

14. Steven Szara in *Psychotropic Drugs*, S. Garattini and V. Ghetti, eds., (Amsterdam: Elsevier, 1957), p. 460.

15. Jay Stevens, *Storming Heaven: LSD and the American Dream* (New York: Atlantic Monthly Press, 1987); Martin A. Lee and Bruce Shlain, *Acid Dreams: The CIA, LSD, and the Sixties Rebellion* (New York: Grove Press, 1985).

16. Lee and Shlain, op. cit., p. xxi.

17. Ibid., p. 286.

18. A. Hoffer and H. Osmond, *New Hope for Alcoholics* (New York: University Books, 1968).

19. Lester Grinspoon and James B. Bakalar, *Psychedelic Drugs Reconsidered* (New York: Basic Books, 1979), p. 216.

20. Lee and Shlain, op. cit., p. 84.

21. Sophia Adamson, *Through the Gateway of the Heart* (San Francisco: Four Trees Press).

15. ANTICIPATING THE ARCHAIC PARADISE

1. Jan G. R. Elferink, "Some Little-Known Hallucinogenic Plants of the Aztecs," *Journal of Psychoactive Drugs* 20, no. 4.

2. Timothy Leary and Ralph Metzner, *The Psychedelic Experience: A Manual Based on the Tibetan Book of the Dead* (New Hyde Park, N.Y.: University Books, 1964).

3. Martin Moyniham, *Communication and Noncommunication by Cephalopods* (Bloomington: Indiana University Press, 1985).

4. Hans Jonas, *The Phenomenon of Life* (New York: Dell, 1966), p. 238.

5. C. G. Jung, *Psychology and Alchemy* (London: Routledge & Kegan Paul, 1953), p. 190.

6. Arnold S. Trebach, *The Great Drug War* (New York: Macmillan, 1987), p. 363.

EPILOGUE: LOOKING OUTWARD AND INWARD TO A SEA OF STARS

1. Arthur Koestler, *The Ghost in the Machine* (New York: Macmillan, 1967), p. 339.

GLOSSARY

Alkaloid: A member of a large class of secondary compounds that contain nitrogen and that usually display one or more kinds of biological activity, such as liver toxicity or effects on the central nervous system. Examples of alkaloids include cocaine, morphine, psilocybin, caffeine, nicotine, and atropine.
Amanita muscaria: The fly agaric, a red-capped, white-spotted mushroom of Siberian shamanism and European folklore that has a symbiotic relationship to birch and fir trees. It was identified with Soma by R. Gordon and Valentina Wasson.
Archaic Revival: The refocusing of public attention on the themes and values of human prehistory. Psychoanalysis, rock and roll, sexual permissiveness, and psychedelic drug use are but a few of the social manifestations of the twentieth century that are arguably part of the Archaic Revival.
Avestan: An ancient Iranian language.
Ayahuasca: A Quechawa word that roughly translates as "vine of the dead" or "vine of souls." This term refers not only to a prepared hallucinogenic beverage, but also to one of the main ingredients of that beverage, the Malpighaecaeous liana, *Banisteriopsis caapi*. This plant, a woody climber, can reach lengths of more than a hundred meters, and a single adult plant can weigh a ton or more. Its tissues, especially the inner cambium of the bark, are rich in alkaloids of the beta-carboline type. The most important beta-carboline occurring in *Banisteriopsis caapi* is harmine.

Beta-carbolines: A subclass of the indole family, some beta-carbolines are hallucinogenic, including harmine, harmaline, tetrahydroharmine, and 6-methoxyharmine.

Bwiti: The Bwiti religion among the Fang of Gabon and Zaire can be called a truly African hallucinogenic plant cult. It is based on the ritual use of the ibogaine-containing root bark of the *Tabernanthe iboga* bush.

Çatal Hüyük: An archaeological site on the Anatolian plain of Asia Minor. Çatal Hüyük has been called "a premature flash of brilliance and complexity" and "an immensely rich and luxurious city." The stratigraphy for the site begins in the middle of the ninth millennium B.C., with elaboration of cultural forms reaching a pinnacle in the middle of the seventh millennium.
Catalysis: A speeding up of processes already occurring, albeit slowly.
Coprophilic: "Dung loving," used to describe species of mushrooms whose preferred environment is the dung of cattle.

Emetic: A purgative, something that causes vomiting.
Endogenous: Occurring within the body as a normal part of metabolism.
Entheogen: A term coined by R. Gordon Wasson, which he preferred to the common term "psychedelic." The word refers to the felt presence of indwelling divinity experienced under the influence of psilocybin.
Epigenetic change: Changes that are not genetic. Learned behaviors such as writing are epigenetic. Books and electronic data bases are epigenetic forms of storage of information. Culture is a learned, hence epigenetic, form.
Ethnomycology: The field founded by R. Gordon and Valentina Wasson. Ethnomycology is the study of human cultural and historical interaction with fungi, especially mushrooms.
Ethnopharmacology: The study of use and preparation of plants and plant drugs in non-Western cultures.
Exogenous: Existing outside of the body, coming from without.
Exopheromones: Chemical messengers that do not act among the members of a single species in the way that insect pheromones have made familiar, but act instead across species lines, thus allowing one species to influence another. Some exopheromones act in ways that allow one species to affect a community of species or an entire biome.

Gaia: The Great Goddess, the horned goddess, mistress of the animals, who is ubiquitous in the art of the Upper Paleolithic. Gaia is popularly equated with Ge, the goddess of the Earth.
Gaian holism: A sense of the unity and balance of nature and of our own human position within that dynamic and evolving balance. It is a plant-based view and a return to a perspective on self and ego that places them within the larger context of planetary life and evolution.
Glossolalia: Spontaneous outbursts of syntactically ordered sound with apparent linguistic intent that sometimes occur during states of religious frenzy or hallucinogen-induced ecstasy.

Haoma: The word for Soma in Zend, the language of the Avestan literature of Zoroasterism.

Heiros gamos: Used in the Jungian sense of an alchemical marriage or a union of opposites that transcends the mundane realm.

Indole hallucinogens: LSD, psilocybin, dimethyltryptamine, ibogaine, and the beta-carbolines are the principal indole hallucinogens (see Figure 28).

Menog: The ordinarily invisible spiritual world of the after-death state according to the *Zend Avesta*.

Mutagen: Something that is the causal agent of mutation. Cosmic rays, toxic chemicals, and some drugs can act as mutagens.

Natufian culture: Middle Eastern culture of 9000 B.C. whose crescent moon flints and elegantly naturalistic carved bonework is unrivaled by anything contemporary found in Europe.

Pandemic: Found worldwide or over a large geographical area.

Partnership: Term introduced by Riane Eisler that refers to a social system in which social relations are primarily based on the principle of linking rather than ranking. In the partnership model, diversity is not equated with either inferiority or superiority. The opposite of this concept is the dominator model. Both matriarchy and patriarchy are considered types of dominator societies.

Pastoralism: A human social style characterized by nomadism and the domestication and husbandry of large animals in a grassland environment. Pastoralists may have partnership arrangements or they may be dominators. The horse-mounted Indo-European pastoralists of the Kurgan waves were certainly dominators. Here I have argued that the Archaic African pastoralism, which was without horses and was based on cattle, was a partnership society.

Peganum harmala: The giant Syrian rue, it grows wild in the drier regions of a range that stretches from Morocco to Manchuria. The plant contains psychoactive indoles of the beta-carboline type.

Pheromones: Chemical compounds exuded by an organism for the purpose of carrying messages between organisms of the same species.

Psilocybin: Hallucinogenically active substance that occurs in the mushroom *Stropharia cubensis* and numerous other species.

Round Head Period: A Tassili-n-Ajjer painting style so named because of the prevalence of depictions of the human figure not known from any other site. The Round Head Period is believed to have begun very early and probably ended before the seventh millennium B.C.

Scythians: A nomadic central Asian barbarian group who entered Eastern Europe around 700 B.C., the Scythians brought the use of cannabis to the European world.

STRUCTURAL TYPES OF PRINCIPAL HALLUCINOGENS

A. Nitrogen Containing (Alkaloidal) Hallucinogens

I. Phenylethylamine Derivatives

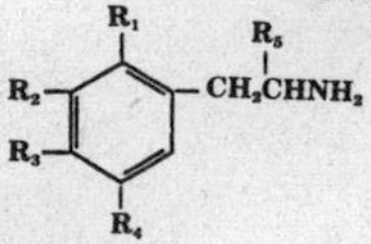

(R_1, R_2, R_3, R_4; CH_2CHNH_2 with R_5)

$R_1 = R_5 = H$, $R_2 = R_3 = R_4 = OCH_3$: 3,4,5-Trimethoxyphenylethylamine (constituent of peyote) = Mescaline

$R_1 = R_4 = OCH_3$, $R_2 = H$, $R_3 = R_5 = CH_3$: 2,5-Dimethoxy-4-methyl-phenylisopropylamine = STP (synthetic compound)

II. Indole Derivatives

1. Tryptamine Derivatives

($CH_2CH_2N(Alkyl)_2$; N–H)

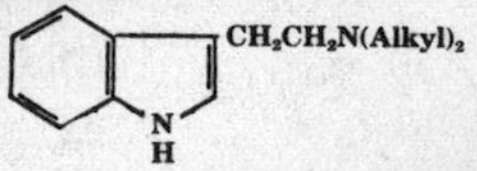

a) $Alkyl = CH_3$: N,N-Dimethyltryptamine (constituent of yopo, etc.)

b) $Alkyl = C_2H_5$, C_3H_7, C_3H_5: N,N,-Diethyltryptamine, etc. (synthetic compounds)

2. Hydroxytryptamine Derivatives

1) 4-Hydroxytryptamine Derivatives:

(R; $CH_2CH_2N(Alkyl)$; N–H)

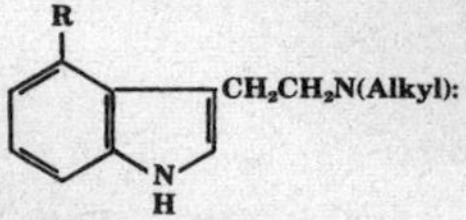

a) $R = OPO_3H$, $Alkyl = CH_3$: Psilocybin
$R = OH$, $Alkyl = CH_3$: Psilocin (constituents of teonanacatl)

b) $R = OPO_3H$, $Alkyl = C_2H_5$: CY-19
$R = OH$, $Alkyl = C_2H_5$: CZ-74
(synthetic compounds)

2) 5-Hydroxytryptamine Derivatives:

(R_1; CH_2CH_2N with R_2, R_3; N–H)

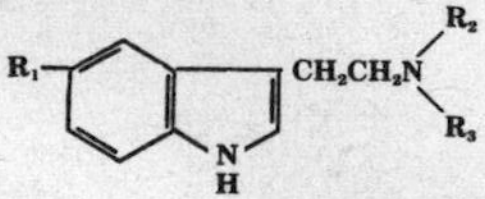

a) $R_1 = OH$, $R_2 = R_3 = CH_3$: Bufotenin

b) $R_1 = OCH_3$, $R_2 = R_3 = CH_3$: 5-Methoxy-N,N-dimethyltryptamine

c) $R_1 = OCH_3$, $R_2 = H$, $R_3 = CH_3$: 5-Methoxy-N-methyltryptamine (constitutents of yopo, coboba, epena, etc.)

3. Cyclic Tryptamine Derivatives

1) β-Carboline Derivatives:

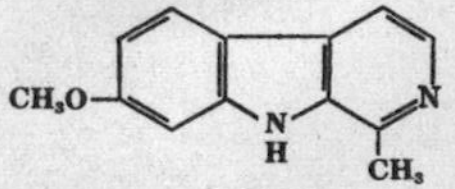

(CH_3O; N–H; N; CH_3)

a) *Harmine*

b) 3,4-Dihydroharmine = Harmaline

c) d-1,2,3,4-Tetrahydroharmine (constituents of ayahuasca, etc.)

2) Lysergic Acid Derivatives:

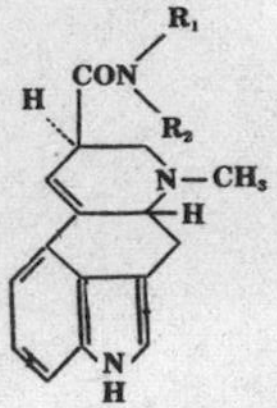

(H; CON with R_1, R_2; N–CH_3; H; N–H)

a) $R_1 = R_2 = H$: d-Lysergic acid amide

b) $R_1 = H$, $R_2 = CHOHCH_3$: d-Lysergic acid hydroxyethylamide (constituents of ololiuqui)

c) $R_1 = R_2 = C_2H_5$: d-Lysergic acid diethylamide (= LSD-25) (synthetic compound)

3) Ibogaine:

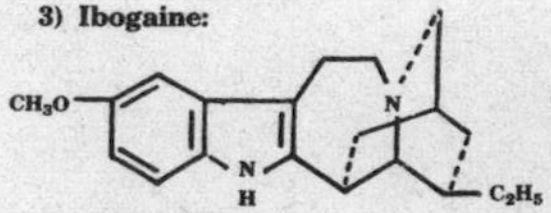

(CH_3O; N–H; N; C_2H_5)

(constitutent of *Tabernanthe iboga*)

FIGURE 28. The Indole Hallucinogens. From *The Invisible Landscape* by Dennis McKenna and Terence McKenna (New York: Seabury Press, 1975), pp. 56–57.

Shamanism: The worldwide tradition of Upper Paleolithic natural magic, it has been beautifully defined by Mircea Eliade as "the archaic techniques of ecstasy." Shamanism continues to be practiced in many parts of the world today.
Stropharia cubensis: Also called *Psilocybe cubensis*, it is the familiar "magic mushroom" grown and loved by mycological and psilocybin enthusiasts worldwide.
Symbiosis: A relationship of mutually productive interdependence among two or more species. A strong symbiotic relationship will result in coevolution of the species involved.

Tabernanthe iboga: A small, yellow-flowered bush that has a history of usage as a hallucinogen in tropical West Africa although it is better known as a powerful aphrodisiac. See **Bwiti**.
Tassili-n-Ajjer Plateau: A curious geological formation in southern Algeria. It is like a labyrinth, a vast badlands of stone escarpments that have been cut by the wind into many perpendicular narrow corridors. Aerial photographs give the eerie impression of an abandoned city. In the Tassili-n-Ajjer are rock paintings that date from the late Neolithic to as recently as two thousand years ago.
Tryptamine hallucinogens: Psilocybin, psilocin, dimethyltryptamine, and their psychoactive structural near relatives.

Virtual reality: Technology currently under development that uses computers, three-dimensional optics, and body imaging to create "virtual environments" in which the user has the impression of being in a real, but alternative, three-dimensional world.

BIBLIOGRAPHY

ADAMSON, Sophia. (1985) *Through the Gateway of the Heart*. San Francisco: Four Trees Press.

ALLEGRO, John M. (1970) *The Sacred Mushroom and the Cross*. Garden City, N.Y.: Doubleday.

AMMERMAN, Albert J. and CAVALLI-SFORZA, Luiga Lucca (1984) *The Neolithic Transition and the Genetics of Populations in Europe*. Princeton: Princeton University Press.

BARTHOLIN, Thomas. (1661) *Historiarum anatomicarum et medicarum rariorum*. Copenhagen.

BEECHING, Jack. (1975) *The Chinese Opium Wars*. New York: Harcourt Brace Jovanovich.

BENDER, Ernest, editor. "R. Gordon Wasson on Soma and Daniel H. H. Ingalls' Response." New Haven: American Oriental Society, Publication #7.

BURROUGHS, William. (1959) *Naked Lunch*. New York: Grove Press.

BURROUGHS, William, and GINSBERG, A. (1963) *The Yagé Letters*. San Francisco: City Lights Books.

COX, George W. (1878) *The Mythology of the Aryan Nations*. London: C. Kegan Paul.

DE QUINCEY, Thomas. (1822) *Confessions of an English Opium-Eater*. London: MacDonald.

DE SEPTCHENES, Le Clerc. (1788) *The Religion of the Ancient Greeks*. London: Elliot and T. Kay.

DICK, Philip K. (1965) *The Man in the High Castle*. London: Penguin.

DOBKIN de RIOS, Marlene. (1972) *Visionary Vine: Psychedelic Healing in the Peruvian Amazon*. San Francisco: Chandler.

DOYLE, Arthur Conan. (1905) *The Complete Sherlock Holmes*. New York: Doubleday.

DROWER, E. S. (1956) *Water into Wine*. London: John Murray.

DUERR, Hans Peter. (1985) *Dreamtime: Concerning the Boundary between Wilderness and Civilization*. Oxford: Basil Blackwell.

DUKE, James A., AULIK, David, and PLOWMAN, Timothy. (1975) "Nutritional Value of Coca." *Botanical Museum Leaflets of Harvard University*, vol. 24, no. 6.

EISLER, Riane. (1987) *The Chalice and the Blade: Our History, Our Future*. San Francisco: Harper & Row.

ELFERINK, Jan G. R. (1988) "Some Little-Known Hallucinogenic Plants of the Aztecs." *Journal of Psychoactive Drugs*, vol. 20, no. 4.

ELIADE, Mircea. (1958) *Yoga: Immortality and Freedom*. New York: Pantheon.

ELIADE, Mircea. (1959) *The Sacred and the Profane*. New York: Harper & Row.

ELIADE, Mircea. (1964) *Shamanism: Archaic Techniques of Ecstasy*. New York: Pantheon.

EMBODEN, William. (1979) *Narcotic Plants*. New York: Macmillan.

FAY, Peter Ward. (1975) *The Opium War*. New York: W. W. Norton.

FERNANDEZ, James W. (1982) *Bwiti: An Ethnography of the Religious Imagination in Africa*. Princeton: Princeton University Press.

FISCHER, Roland, HILL, Richard, THATCHER, Karen, and SCHLIEB, James. (1970) "Psilocybin-Induced Contraction of Nearby Visual Space." *Agents and Actions*, vol. 1, no. 4.

FLATTERY, David, and SCHWARTZ, Martin. (1989) *Haoma and Harmaline*. Near Eastern Studies, vol. 21. Berkeley: University of California Press.

FREUD, Sigmund. (1963) *The Cocaine Papers*. Vienna: Dunquin Press.

FURST, Peter. (1978) *Hallucinogens and Culture*. San Francisco: Chandler & Sharp.

GARROD, D. A. E. (1957) *The Natufian Culture: The Life and Economy of a Mesolithic People in the Near East*. Proceedings of the British Academy 43: 211–227.

GETTINGS, Fred. (1981) *Dictionary of Occult, Hermetic, and Alchemical Sigils*. London: Routledge & Kegan Paul.

GIMBUTAS, Marija. (1982) *The Goddesses and Gods of Old Europe, 6500–3500 B.C.: Myths and Cult Images*. Berkeley: University of California Press.

GRACIE, and ZARKOV. (1986) "An Indo-European Plant Teacher." *Notes from Underground* 10, Berkeley.

GRAVES, Robert. (1948) *The White Goddess*. New York: Creative Age Press.

GRAVES, Robert. (1955) *The Greek Myths*. 2 vols. Baltimore: Penguin.

GRAVES, Robert. (1960) *Food for Centaurs*. Garden City, N.Y.: Doubleday.

GRAVES, Robert. (1964) *Difficult Questions, Easy Answers*. Garden City, N.Y.: Doubleday.

GRINSPOON, Lester, and BAKALAR, James B. (1979) *Psychedelic Drugs Reconsidered*. New York: Basic Books.

GRISWOLD, H. D. (1923) *The Religion of the Rigveda*. London: Oxford University Press.

GROF, Stanislav. (1980) *LSD Psychotherapy*. Pomona, CA: Hunter House.
GROF, Stanislav. (1985) *Beyond the Brain: Birth, Death, and Transcendence in Psychotherapy*. New York: State University of New York Press.
GUENTHER, Herbert V. (1966) *Tibetan Buddhism without Mystification*. Leiden, Netherlands: E. J. Brill.
HAYTER, Alethea. (1968) *Opium and the Romantic Imagination*. Berkeley: University of California Press.
HENRY, Donald Owen. (1973) *The Natufian of Palestine: Its Material Culture and Ecology*. Ann Arbor: University Microfilms.
HERER, Jack. (1990) *The Emperor Wears No Clothes*. Van Nuys, CA: Hemp Publishing.
HERODOTUS. (1901) *Works*. H. Cary, trans. London: George Bell and Sons.
HOBHOUSE, Henry. (1985) *Seeds of Change: Five Plants That Transformed Mankind*. New York: Harper & Row.
HOFFER, A., and OSMOND, H. (1967) *The Hallucinogens*. New York: Academic Press.
HOFFER, A., and OSMOND, H. (1968) *New Hope for Alcoholics*. New York: University Books.
HOFMANN, Albert. (1983) *LSD My Problem Child*. Los Angeles: Tarcher.
HOOKE, S. H. (1963) *Babylonian and Assyrian Religion*. Norman: University of Oklahoma Press.
HUXLEY, Aldous. (1954) *The Doors of Perception*. New York: Harper.
HUXLEY, Aldous. (1977) *Moksha: Writings on Psychedelics and the Visionary Experience*. Michael Horowitz and Cynthia Palmer, eds. New York: Stonehill.
JACOBS, Barry L. (1984) *Hallucinogens: Neurochemical, Behavioral, and Clinical Perspectives*. New York: Raven Press.
JARRY, Alfred. (1965) *Selected Works of Alfred Jarry*. Roger Shattuck and Simon Watson Taylor, eds. New York: Grove Press.
JAYNES, Julian. (1977) *The Origin of Consciousness in the Breakdown of the Bicameral Mind*. Boston: Houghton Mifflin.
JINDRAK, K. F., and JINDRAK, H. (1988) "Mechanical Effect of Vocalization of Human Brain and Meninges," in *Medical Hypotheses*, 25, pp. 17–20.
JONAS, Hans. (1966) *The Phenomenon of Life*. New York: Dell.
JUDD, Elizabeth. (1980) "Hallucinogens and the Origin of Language." *Sociolinguistic Newsletter*, vol. ii, pp. 7–12.
JUNG, C. G. (1953) *Psychology and Alchemy*. London: Routledge & Kegan Paul.
KAEGI, Adolf. (1886) *The Rig Veda: The Oldest Literature of the Indians*. Boston: Ginn.
KERÉNYI, Carl (1976) *Dionysos: Archetypal Image of Indestructible Life*. Bollingen Series LXV. Princeton: Princeton University Press.
KLÜVER, Heinrich. (1928) *Mescal, the Divine Plant and Its Psychological Effects*. London: Kegan Paul.
KOESTLER, Arthur. (1967) *The Ghost in the Machine*. New York: Macmillan.
KRIPPNER, S., and DAVIDSON, R. (1974) "Paranormal Events Occurring during Chemically Induced Psychedelic Experience and Their Implications for Religion." *Journal of Altered States of Consciousness* 1:175.

KRÜGER, Henrik. (1980) *The Great Heroin Coup: Drugs, Intelligence, and International Fascism*. Boston: South End Press.

LA BARRE, Weston. (1972) *The Ghost Dane: Origins of Religion*. New York: Delta Press.

LA WALL, Charles H. (1927) *The Curious Lore of Drugs and Medicines through the Ages*. Philadelphia: J. B. Lippincott.

LACROIX, J. F. de. (1772) *Anecdotes Arabes et Musulmanes, Depuis l'An de J. C. 614*. Paris: Vincent.

LAJOUX, Jean-Dominique. (1963) *The Rock Paintings of Tassili*. Cleveland: World.

LEARY, Timothy, and METZNER, Ralph. (1964) *The Psychedelic Experience: A Manual Based on the Tibetan Book of the Dead*. New Hyde Park, N.Y.: University Books.

LEE, Martin A., and SHLAIN, Bruce. (1985) *Acid Dreams: The CIA, LSD, and the Sixties Rebellion*. New York: Grove Press.

LEWIN, Lewis (1931) *Phantastica: Narcotic and Stimulating Drugs*. New York: E. P. Dutton.

LEWIN, Roger. (1988) *In the Age of Mankind*. New York: Smithsonian Institution.

LEWIN, Roger. (1991) "Stone Age Psychedelia." *New Scientist*. 8 June, pp. 30–34.

LHOTE, Henri. (1959) *The Search for the Tassili Frescoes*. New York: E. P. Dutton.

LUDLOW, Fitz Hugh. (1857) *The Hasheesh Eater: Being Passages from the Life of a Pythagorean*. New York: Harper & Brothers.

LUMSDEN, Charles J., and WILSON, Edward O. (1983) *Promethean Fire: Reflections on the Origin of Mind*. Cambridge, Mass.: Harvard University Press.

LUNA, Luis Eduardo. (1986) *Vegetalismo: Shamanism among the Mestizo Population of the Peruvian Amazon*. Stockholm: Almquist & Wiksell.

LUNA, Luis Eduardo, and AMARINGO, Pablo. (1991) *Ayahuasca Visions: The Religious Iconography of a Peruvian Shaman*. Berkeley: North Atlantic Books.

MANDER, Jerry. (1978) *Four Arguments for the Elimination of Television*. New York: Quill.

MARCHETTI, Victor, and MARKS, John D. (1974) *The CIA and the Cult of Intelligence*. New York: Albert A. Knopf.

MASPERO, Gaston. (1894) *The Dawn of Civilization: Egypt and Chaldea*. London: Society for Promoting Christian Knowledge.

MATHESON, Richard R. (1958) *The Eternal Search: The Story of Man and His Drugs*. New York: G. P. Putnam's Sons.

MATOSSIAN, Mary Allerton Kilbourne. (1989) *Poisons of the Past: Molds, Epidemics, and History*. New Haven: Yale University Press.

MCCOY, Alfred W. (1972) *The Politics of Heroin in Southeast Asia*. New York: Harper Colophon Books.

MCKENNA, Dennis, and MCKENNA, Terence. (1975) *The Invisible Landscape*. New York: Seabury Press.

MCKENNA, Dennis, TOWERS, G. H. N., and ABBOTT, F. S. (1984a) "Monoamine Oxidase Inhibitors in South American Hallucinogenic Plants, Part I: Tryptamine and Beta-carboline Constituents of Ayahuasca." *Journal of Ethnopharmacology* 10:195–223.

MCKENNA, Dennis, TOWERS, G. H. N., and ABBOTT, F. S. (1984b) "Monoamine Oxidase Inhibitors in South American Hallucinogenic Plants, Part II: Constituents of Orally-Active Myristicaceous Hallucinogens." *Journal of Ethnopharmacology* 12:179–211.

MELLAART, James. (1965) *Earliest Civilizations of the Near East*. New York: McGraw-Hill.

MELLAART, James. (1967) *Çatal Hüyük: A Neolithic Town in Anatolia*. New York: McGraw-Hill.

MILLER, Jean Baker. (1986) *Toward a New Psychology of Women*. Boston: Beacon Press.

MOYNIHAM, Martin. (1985) *Communication and Noncommunication by Cephalopods*. Bloomington: Indiana University Press.

MORTIMER, W. Golden. (1974) *History of Coca: The Divine Plant of the Incas*. San Francisco: Fitz Hugh Ludlow Library Edition.

MUNN, Henry. (1973) "The Mushrooms of Language." In Michael J. Harner, ed., *Shamanism and Hallucinogens*. London: Oxford University Press.

MYLONAS, George E. (1961) *Eleusis and the Eleusinian Mysteries*. Princeton: Princeton University Press.

NARANJO, Claudio. (1973) *The Healing Journey: New Approaches to Consciousness*. New York: Ballantine.

NILSSON, Martin P. (1964) *A History of Greek Religion*. New York: W. W. Norton.

NEUMANN, Erich. (1955) *The Great Mother: An Analysis of the Archetype*. New York: Pantheon.

O'LEARY, De Lacy. (1949) *How Greek Science Passed to the Arabs*. London: Routledge & Kegan Paul.

OMAN, J. Campbell. (1903) *The Mystics, Ascetics, and Saints of India*. London: T. Fischer Unwin.

OSS, O. T., and OERIC, O. N. (1976) *Psilocybin: The Magic Mushroom Grower's Guide*. Berkeley: Lux Natura Press.

OTT, Jonathan. (1985) *The Cacahuatl Eater: Ruminations of an Unabashed Chocolate Eater*. Vashon, W.A.: Natural Products Co.

OTT, Jonathan, and BIGWOOD, Jeremy, eds. (1978) *Teonanacatl Hallucinogenic Mushrooms of North America*. Seattle: Madrona Publishers.

OTTO, Walter F. (1955) "The Meaning of the Eleusinian Mysteries." In Joseph Campbell, *Eranos Yearbook Number Two: The Mysteries*. New York: Pantheon.

OTTO, Walter F. (1965) *Dionysus Myth and Cult*. Bloomington: Indiana University Press.

PERSSON, Axel W. (1942) *The Religion of Greece in Prehistoric Times*. Berkeley: University of California Press.

PFEIFFER, John E. (1982) *The Creative Explosion: An Inquiry into the Origins of Art and Religion*. Ithaca, N.Y.: Cornell University Press.

PHELPS, Janice Keller, and NOURSE, Alan E. (1986) *The Hidden Addiction and How to Get Free.* Boston: Little, Brown.

RÄTSCH, Christian. (1984) *Ein Kosmos im Regenwald.* Cologne: Eugen Diederichs Verlag.

RÄTSCH, Christian. (1986) *Ethnopharmakilogie und Parapsychologie.* Berlin: Express Edition GmbH.

RÄTSCH, Christian, and MÜLLER-EBELING, Claudia. (1986) *Isoldens Liebestrank Aphrodisiaka in Geschichte und Gegenwart.* Munich: Kindler Verlag.

REKO, Victor A. (1987) *Magische Gife, Rausch-und Betäubungsmittel der neuen Welt.* Berlin: Express Edition GmbH.

RENFREW, A. Colin. (1988) *Archaeology and Language: The Puzzle of Indo-European Origins.* London: Cambridge University Press.

ROBICSEK, Francis. (1978) *The Smoking Gods: Tobacco in Maya Art, History and Religion.* Norman: University of Oklahoma Press.

RODRIGUEZ, E., AREGULLIN, M., UEHARA, S., NISHIDA, T., WRANGHAM, R., ABRAMOWSKI, Z., FINLAYSON, A., and TOWERS, G. H. N. (1985) "Thiarubrine-A, A Bioactive Constituent of *Aspilia (Asteraceae)* Consumed by Wild Chimpanzees." *Experientia* 41:419–420.

RULANDUS, Martinus. (1612) *A Lexicon of Alchemy or Alchemical Dictionary.* Frankfurt: Zachariah Palthenus.

SAMORINI, Giorgio. (1989) "Etnomicologia nell'arte rupestre Sahariana (Periodo delle 'Teste Rotonde')." *Boll. Camuno Notizie,* vol. 6(2):18–22.

SAUR, Carl. (1973) *Man's Impact on the Earth.* New York: Academic Press.

SCHULTES, Richard Evans. (1939) "Plantae Mexicanae, II: The Identification of Teonanácatl, a Narcotic Basidiomycete of the Aztecs." *Botanical Museum Leaflets of Harvard University* 7:37–54.

SCHULTES, Richard Evans. (1973) *The Botany and Chemistry of Hallucinogens.* Springfield, Mass.: Charles C. Thomas.

SCHULTES, Richard Evans. (1982) "The Beta-Carboline Hallucinogens of South America." *Journal of Psychoactive Drugs* 14:205–220.

SCHULTES, Richard Evans, and RAFFAUF, Robert F. (1990) *The Healing Forest: Medicinal and Toxic Plants of Northwest Amazonia.* Portland, OR: Dioscorides Press.

SETTEGAST, Mary. (1987) *Plato Prehistorian.* Cambridge: Rotenberg Press.

SHEVOROSHKIN, Vitaly. (1990) "The Mother Tongue." *The Sciences,* May/June, pp. 20–27.

SIEGEL, Ronald K. (1977) "Religious Behavior in Animals and Man: Drug-Induced Effects." *Journal of Drug Issues,* pp. 219–236.

SIEGEL, Ronald K. (1989) *Intoxication.* New York: E. P. Dutton.

SPRUCE, Richard. (1908) *Notes of a Botanist on the Amazon and Andes* (2 vols.). A. R. Wallace, ed.

STAHL, Peter W. (1989) "Identification of Hallucinatory Themes in the Late Neolithic Art of Hungary." *Journal of Psychoactive Drugs* 21(1):101–112

STEVENS, Jay. (1987) *Storming Heaven: LSD and the American Dream.* New York: Atlantic Monthly Press.

STEVENS, Wallace. (1981) *The Collected Poems of Wallace Stevens.* New York: Alfred A. Knopf.

SUHR, Elmer G. (1967) *Before Olympos.* New York: Helios Books.

SZARA, Steven. (1957) *Psychotropic Drugs.* S. Garattini and V. Ghetti, eds., Amsterdam: Elsevier.

TAYLOR, Bayard. (1855) *The Lands of the Saracen.* New York: G. P. Putnam.

TREBACH, Arnold S. (1987) *The Great Drug War.* New York: Macmillan.

VARELA, Francisco J., and COUTINHO, A. (1988) "The Body Thinks: How and Why the Immune System Is Cognitive." *The Reality Club*, vol. 2, John Brickman, ed. New York: Phoenix Press.

WADDINGTON, C. H. (1961) *The Nature of Life.* London: George Allen & Unwin.

WALEY, Arthur. (1958) *The Opium War Through Chinese Eyes.* Stanford: Stanford University Press.

WALTON, Robert P. (1938) *Marijuana: America's New Drug Problem.* Philadelphia: J. B. Lippincott.

WARE, James R. (1966) *Alchemy, Medicine, Religion in the China of A.D. 320: The Nei Pien of Ko Hung.* Cambridge, Mass.: MIT Press.

WASSON, R. Gordon. (1971) *Soma: Divine Mushroom of Immortality.* New York: Harcourt Brace Jovanovich.

WASSON, R. Gordon. (1980) *The Wondrous Mushroom: Mycolatry in Mesoamerica.* New York: McGraw-Hill.

WASSON, R. Gordon. (1986) *Persephone's Quest: Entheogens and the Origins of Religion.* New Haven: Yale University Press.

WASSON, R. Gordon, and HEIM, Roger. (1958) *Les Champignons Hallucinogènes du Mexique.* Paris: Editions du Musée National d'Histoire Naturelle.

WASSON, R. Gordon, and HEIM, Roger. (1967) *Nouvelles Investigations sur les Champignons Hallucinogènes.* Paris: Editions du Musée National d'Histoire Naturelle.

WASSON, R. Gordon, HOFMANN, Albert, and RUCK, Carl. (1978) *The Road to Eleusis.* New York: Harcourt Brace Jovanovich.

WILSON, Edward O. (1984) *Biophilia.* Cambridge, Mass.: Harvard University Press.

WILSON, H. H., trans. (1928) *Rig-Veda Sanhita* (5 vols.). Poona, India: Ashtekar.

WINDHOLZ, Martha, ed. (1976) *The Merck Index* (9th ed.). Rahway, N.J.: Merck.

WINN, Marie. (1977) *The Plug-In Drug.* New York: Penguin.

ZAHNER, R. C. (1961) *The Dawn and Twilight of Zoroastrianism.* New York: G. P. Putnam's Sons.

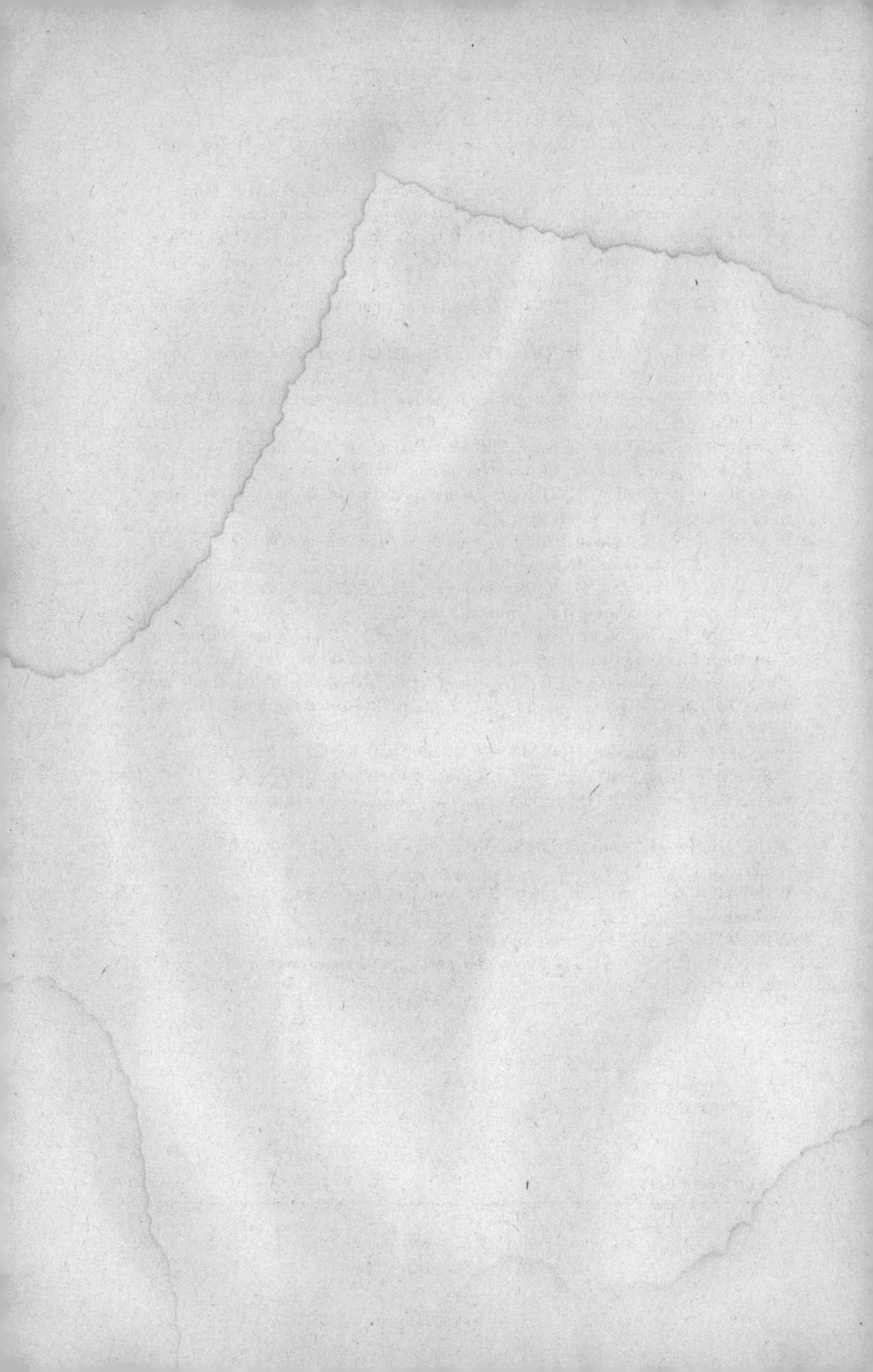

INDEX

Page numbers of diagrams and illustrations appear in italics.

Grateful acknowledgment is made for permission to reprint the following:

Excerpt from *Seeds of Change: Five Plants That Changed the World* by Henry Hobhouse. Copyright © 1985 by Henry Hobhouse. Reprinted by permission of HarperCollins Publishers.

Excerpt from *Water into Wine* by E. S. Drower. Reprinted by permission of John Murray Publishers Ltd.

Excerpt from *The Chalice and the Blade* by Riane Eisler. Copyright © 1987 by Riane Eisler. Reprinted by permission of HarperCollins Publishers.

Excerpt from *Narcotic Plants* by William Emboden. Copyright © 1979 by William Emboden, Jr. Reprinted by permission of Macmillan Publishing Company.

Excerpt from *History of Coca: Divine Plant of the Incas* by W. G. Mortimer. Reprinted by permission.

Excerpt from *Acid Dreams* by Martin Lee and Bruce Shlain. Reprinted by permission.

Excerpt from *The Plug-In Drug* by Marie Winn. Copyright © 1977, 1985 by Marie Winn Miller. Reprinted by permission of Viking Penguin, a division of Penguin Books USA Inc.

Excerpt from *The Great Drug War* by Arnold Trebach. Reprinted by permission of Dominick Abel Literary Agency, Inc.

Excerpt from *Intoxication* by Ronald K. Siegel. Copyright © 1989 by Ronald K. Siegel, Ph.D., Inc. Reprinted by permission of the publisher, Dutton, an imprint of New American Library, a division of Penguin Books USA Inc.

Excerpt from *Dreamtime* by Peter Duerr. Reprinted by permission of Basil Blackwell Ltd.

Excerpt from *The Hallucinogens* by A. Joffer and H. Osmond. Reprinted by permission.

Excerpt from *The Natufian of Palestine* by Dr. Donald Owen Henry. Reprinted by permission.

ABOUT THE AUTHOR

Terence McKenna, author and explorer, has traveled the world to work and live with shamans. He has added to their shared knowledge of rituals his own efforts to preserve the plants used in these ceremonies. Co-author of *The Invisible Landscape* and *Psilocybin: The Magic Mushroom Grower's Guide,* Terence mesmerizes his many lecture audiences with tales of science and shamanism. He lives in Occidental, California, and is co-manager of a botanical garden in Hawaii for endangered tropical plants.